I0485941

Problem Solving in Solid Waste Engineering

By

Prof. Dr. Eng. Isam Mohammed Abdel-Magid Ahmed

and
Dr. Mohammed Isam Mohammed Abdel-Magid

Second edition reviewed, revised, updated and improved, 2015

First edition printed by University of Dammam Press, Dammam, 2012.

ISBN-13: 978-1515378976
ISBN-10: 1515378977

© Authors.

Prof. Dr. Eng. Isam Mohammed Abdel-Magid Ahmed: Head Proofreading and revision department at the Centre of Scientific Publications and Dammam University, Professor of water resources and environmental engineering, Building 800, Room 240 Environmental Engineering Department, College of Engineering, University of Dammam, Box 1982, Dammam 31451, KSA, Fax: +96638584331,   Phone: +966530310018,   E-mail: iahmed@ud.edu.sa,   isam_abdelmagid@yahoo.com,   Web   site: http://www/sites.google.com/site/isamabdelmagid/

Dr. Mohammed Isam Mohammed Abdel-Magid: Department of Internal Medicine, Khasab Hospital & Polyclinic, P. O. Box 306, Postal 811, Khasab, Musandam, Sultanate of OmanPhone: +97470445235,   +971558215655,   +249969263307,   E-mail: mohammed_isam1984@yahoo.com,   website: http://sites.google.com/site/mohammedisam2000

Preface to Second Edition

"Problems solving in solid waste engineering" is the first and only unique book that address problem solving solid waste issues from engineering, mathematical and computer modeled format and perspective. The book provides basic logging of computer programs as they are used to survey solid waste characteristics, collection and treatment systems. Powerful computer programs are used at every stage and section of the various chapters of the book to aid understanding and employment of fundamental concepts and approaches.

The first version of the book was launched to environmental engineering candidates at college of engineering, University of Dammam. The book is mainly used as a course supplement and support in problem solving issues. Constructive comments, valuable remarks, precious notes and helpful observations were received from students, users within the college, colleagues, engineers, officials at solid waste departments and municipalities, members of engineering societies and enterprises. This accomplishment and success motivated the authors to revise the book, update its contents, improve settings and add more valuable problems to upgrade the version. In this second issue problem modeling techniques has been introduced. Visual Basic.NET, programmed using Microsoft Visual Studio 2010 IDE was used in writing computer programs for selected examples in the book. Set programs are constructed using the IDE designing and buildings tools, and were tested and run on a MS-Windows XP and 7 workstations.

The numbers of individuals, esteemed friends, colleague and many others who have made existence of this text possible are numerous. At the hazard of leaving someone out, the authors would like to thank all individuals for their valuable contributions, many helpful comments, and useful suggestions in bringing out editions of the book. Sincere thanks are extended to the many thoughtful students[1] who helped to solve problems, extract useful information, share academic resources, point out online knowledge, proofread text, prepare illustrations, raise embarrassing questions, and aid other students understand subject area.

As always, the authors appreciate any comments, explanations, observations, notes, suggestions, corrections, and contributions for future and forthcoming revisions of this popular book.

Prof. Dr. Eng. Isam Mohammed Abdel-Magid Ahmed
Dr. Mohammed Isam Mohammed Abdel-Magid

Alkhobar, Khasab, 2015

---

[1] Ibrahim Balgeeth Albargi, Abdulaziz Khalid AlGusaibi, Khalid Mohammed Abdullah Ahmed, Abduaziz Tariq Alasoum, Abdulilah Gafar Habeeb Alhadad, Dawood Mustafa Alqasab, Mohammed Waleed Rashid AlTihani, Mohammed Farhan AlKhaldi, Abdulelah Khalid Aldiruosh, Abdulrahman Ahmed Alhawas, Abdulazia Salih AlGabr, Abdulrahman Muaz Alkhunaini, Mohammed Salih Algarawi, Mufeed Rida Alabkari, Baqer Amin Alnasir, Ibrahim Yonos Alkhaldi, Mohammed Abdulaziz Alzarah, Mutaz Abdulrahman Alhadlag, Abdulrahman Hassan Alqadi, Abdulsalam Mohammed AlKhalaf, Fahad Khalid Alrashodi, Saleh Hamdan, Fahad Dhafir Alyami, Basil Hassan Alzufairi, Almutasimbillah Yahya Asiri, Mamdouh Muflih Alanzi, Abdulrahamn Daij Aldaij

4

Preface to First Edition

Many books and text material are to be found addressing solid waste generation sources, methods of collection, treatment technology, reuse and final disposal. Nevertheless, rather very few, if any, books deal with practical problems of municipal solid waste and solutions.

This book has been written to cover material outlined in the 3-credit hour course of "solid and hazardous waste management, ENV-462" for senior fourth year engineering students of the Environmental Engineering Department of the College Of Engineering At Dammam University. The named course contents included: Types and sources of solid waste; chemical and physical properties of municipal and industrial refuse; solid waste collection methods; solid waste treatment and disposal techniques with emphasis on: landfill disposal, incineration, composting and pyrolysis; salvage, reclaiming, and recycle operations; economics of disposal methods; advantages and disadvantages of each; special and hazardous waste handling; operation and management of solid and hazardous waste programs.

Objectives of the book are intended to fulfill the main learning outcomes for students enrolled in the course, which included the following:

- Enlighten the state of the art in technology, organizational and legislative developments and practices of handling solid wastes.
- Discussing in depth financial burdens of municipal solid waste and benefits as well as solutions.
- Help students to make informed decisions in their professional activities. Thus, assisting them in defining and implementing integrated solutions to the challenges posed by solid wastes in an urban environment.
- Support students to establish a solid waste management, SWM, system that is capable of functioning not only in situations where sufficient resources are available but also

under the more challenging conditions usually prevailing in large cities in low-income countries.

- Acquisition of knowledge by learning new concepts in solid waste management, final disposal, recycling and reuse.
- Cognitive skills through critical thinking and problem solving.
- Numerical skills through application of knowledge in basic solid waste kinetics and mathematical models.
- Student becomes responsible for their own learning through solution of assignments, home works, exercises and report writing.

Knowledge to be acquired from the course is expected to incorporate:

- Quantifying and characterizing MSW in the Kingdom of Saudi Arabia, KSA, and to understand properties commonly associated with MSW.
- Describing tasks and logistics of MSW collection, to analyze collection systems, and to become familiar with principals and theory behind the use of transfer stations.
- Describing the components of a sanitary landfill and processes which take place in a landfill aerobic reactor.
- Doing basic design calculations associated with MSW systems.
- Describing the most common waste processing techniques and their application areas.
- Identifying concepts of MSW reduction, reuse and recycling.
- Develop a strategy to deal with special wastes, hazardous and health care wastes.
- Knowledge of behavior and properties of solid wastes and concepts of engineering control and treatment principles.
- Collection of data, analysis and interpretation.

Cognitive skills to be developed are expected to incorporate the following:

6

- Capturing ability of reasonable scientific judgment and concepts of appropriate decision making.
- Students will be able to apply the knowledge of behavior and properties of solid waste materials that they have learnt in this course in a practical environmental engineering domain.
- Students should be able to design and apply necessary procedures and precautions to produce durable and environmentally usable solid waste products.
- Students will be able to understand the use and application of control materials and components in solid waste collection, treatment, disposal and integrated management.

Solid waste and garbage disposal constitutes a huge responsibility for the producer, community and governmental municipality or authority responsible for its collection, sorting, treatment and final disposal. If the concerned authority did not have good management for its disposal, it exposes itself to environmental, political, economic and social problems. This is due to many interrelated factors such as:

- Quantity produced
  - o Garbage piles up on roads, streets and parks producing foul odors.
- Social disturbances
  - o MSW causes great inconvenience to residents, neighboring area and surroundings.
  - o Hurt sightseeing.
  - o Expel tourists and limit their entry to the region.
  - o Disturbs peace and coexistence of collective tribal in rural areas, villages and cities.

  - o Unpleasant and undesirable odors resulting from bacterial decomposition of organic materials of components of the waste.

- Health problems
  - o Harm workers and patrons.
  - o Introduces sickness to children and animals.

7

- o Breeding areas for rats, mice, vermin, fleas and flies and vectors that transmit infectious diseases, and outbreaks of diseases such as plague.

- o Endemic and epidemic, raising rates of death, destruction, and loss of crops and cattle.

- o Direct and indirect spread of epidemics and diseases through waste accumulation in settlements (plague, malaria, dengue fever, typhus, cholera).

- o Burning of solid waste leads to increased air pollution and respiratory diseases.
- Engineering malfunctions
  - o Disrupt traffic.
  - o Risk of flooding and contamination of water resources.

- Environmental concerns
  - o Introduces pollutants and contaminants to waterways.
  - o Ill environmental impact and pollution of water, air and soil.

  - o Environmental pollution resulting from household waste, street cleaning and dumps areas .... etc..

  - o Bio-chemical and microbiological pollution of ground and surface water due to unregulated disposal of waste.

  - o Air pollution and the presence of organic and inorganic toxic materials, especially in industrial solid waste.

"Problem solving in solid waste engineering" is primarily designed as a supplement and a complementary guide to municipal solid waste engineering. Nonetheless, it can be used as an independent problem

8

solving text in solid waste collection, treatment and disposal. The book targets university students and solid waste engineering candidates taking first degree courses in environmental, civil, mechanical, construction and chemical engineering or related fields. The manuscript is expected to be of beneficial use to postgraduate students and professional engineers. Likewise, it is hoped that the book will stimulate problem solving learning and facilitate self-teaching. By writing such a script it is hoped that the included worked examples and problems will ensure that the booklet is a valuable aid to student-centered learning. To achieve such objectives immense care was taken to present solutions to selected problems in a clear and distinct format using step-by-step procedure and explanation of the related solution utilizing necessary methods, approaches, equations, data, figures and calculations.

The author is grateful to all those authors, writers, researchers and scientists whose methods, procedures, techniques and models have been used throughout the text for solving worked examples. The author acknowledges the inspiration, motivation and stimulus help offered to him by Dean Dr. Abdul-Rahman ben Salih Hariri, Dean College of Engineering of University of Dammam. The author is very fortunate, blessed and thankful to have had such[2] a bright, smart and dedicated group and class of students. Their existence, enthusiasm and dedication inspired teaching. Besides, it has been a real pleasure and enjoyment advising and mentoring such fine young

---

[2] Abdulaziz Al-Abdulkarim, Abdullah Al-Hamed, Abdulaziz Al-Qahtani, Abdulaziz Al-Ghawinem, Abdullah Al-Khamis, Ahmed Al-Safi, Ibrahim Al-Rashid, Ibrahim Al-Shaygi, Mohammed Al-Hosan, Mohamed Al-Khadra, Muhammed Al-Rayaan, Naif Al-Qahtani, Saad Al-Qahtani, Shaheen Al-Dossary, Sultan Al-Rubaish.
   Abdulazeez Fahad Almuhanna, Ahmed Abbas Ibrahim Alshamri, Ahmed Abdullah Salih Aldosari, Ammar Rashed Aldosari, Badr Saleh Alzahrani, Bandar Abdulrahman Mukhtar Makhdom, Fahad Abdulkareem Tahir Alhamad, Hasan Abdrbalrsool Alsadig, Ibrahim Mohammed Aldaiyouli, Khalid Saad Muslih Alshehri, Mohammed Ibrahim Ali Alnabbat, Mohammed Ahmed Alyhyai, Mohammed Shukri Mohammed Abdulrazig, Mohammed Nabeel Mohammed Alghamdi, Muath Khalid Hamad Alfaraj, Omer Salih Alghamdi, Tariq Hadi Alfaifi, Talal Fahad Alotaibi and Waleed Abdulrazag Shukri

promising future scientists. Special and sincere vote of thanks would go to Mr. Mugbil ben Abdullahi Al-Rowais the director of Dammam University Press and his supreme technical staff for the neat typing of the book.

Prof. Dr. Eng. Isam Mohammed Abdel-Magid Ahmed
Professor of water resources and environmental engineering
Al-Thugba, AlKhobar, October 2012

Table of contents

List of Figures

List of Tables

List of Computer Programs

List of Appendixes

Chapter One

Computer Models used in the book

1.1 Introduction and general design

Computers were once considered the future of information, but now they are part of every day activity. The abstract idea of computation involves so many things in common use today, starting – but not stopping – at a conventional full-sized desktop computer. Running through laptops, notebooks, tablets, smartphones, and even smart-watches, computational theory and practice spans a whole spectrum of devices, some are general in purpose; some are specialized, even highly specialized like an embedded Linux or Windows operating system running on a modern wrist-watch.

Using computer in engineering modeling is commonplace today. This is to the extent that practicing professional engineering without working knowledge of, at least, the basic computer modeling methods is, to the least, considered a major drawback in practice.

There are multiple computer models, with different implementation methods. These methods differ according to the underlying programming language used (be it a general purpose or a specialized programming language), the computer platform (most commonly used in market being the Intel-compatible x86 or AMD64 architectures), the operating system workings (again, common OSes in the market including – but not limited to – MS-Windows, Linux with all its flavors, MAC OS, UNIX, among others), and above all, the engineering backbone used in modular design.

Nonetheless, using computers in modeling is not an easy task. An engineer modeler teaming with a programmer, or even better, an engineer programmer by himself, has many angels to consider in model design:

14

- What kind of architecture is the model intended to run at? Is it for desktop PCs? For embedded operating systems? To be run under 32-bit or 64-bit compatible processors? To work under single-processing or multi-processing environment? ... Most general purpose engineering models, being common and easy to run with no special hardware needs, will be run under one or another Intel-compatible processor, running on an IBM-compatible PC or laptop, under 64-bit (supporting 32-bit) single or multi-processing environment (not to mention multithreading programs).

- What kind of operating system is it going to support? This depends not only on the programmer's preference and experience, but also is enforced by the audience who will run and use the software. There is no point in writing Windows native executable programs for a company that runs Linux based computers, for example. This specific example may look superficial, but there are many other examples in living programming modeling that are much more complicated, but the same idea holds.

- Based on the last point, cross-platform support is becoming an important issue in software engineering. To reach the widest possible group of audience, programmers must consider writing code to be run under the largest possible (or at least the most common) types of operating systems. This can be achieved by writing code in a cross-platform language that is run by an interpreter or an on-the-fly compiler (like Java or the .NET framework languages). The hard way to go is to write the code for one OS, migrate it (which needs code revision and/or editing) to another platform, using OS-native tools to recompile and rebuild the program. This way is harder, takes longer time, is error-prone, and every change in code will need to be mirrored into all the other versions, but is guaranteed to end with executables that are native to each operating system, are faster and more reliable.

- What programming language is going to be used? Some general programs need general purpose languages like VB, VC++, C/C++ among others. Programming for cross-

15

platforming may involve other languages like Java or Qt/C++. Graphical detailing, mathematical modeling, web programming, database access, all need special purpose programming languages that may(not) need their specialized compilers/interpreters to run effectively.

That being settled, the programmer will need to put a general blueprint for the program flow, which is commonly known as the program's algorithm. This will involve answering questions such as:

- What is the intended purpose of the program? E.g., "A program to calculate the square root of a number"
- What kind of inputs does it accept? E.g., "program accepts one floating point number"
- What are the steps of computations needed to be done, including any equations or special mathematical formulae that will be used? E.g., "calculate the square root using sqrt function"
- What are the outputs expected from the program? E.g., "prints the sqrt of the input"

This may(not) be supported by a visual layout or a "flow chart". For the sake of simplicity, all programs in this book included only the textual representation of each program's algorithm.

1.2 Programming examples

This book is not intended to be an introduction to computer programming, nor is it going to dive into details about computer architecture or operating system workings. The reader is assumed to have a basic working knowledge in the following areas to be able to read, follow, test, and implement the following programming examples given through this text:

- Basic computer knowledge and how-to's (as how to open/operate/setup a computer workstation),
- Installing and running any operating system (computer programs in the book are programmed using Visual Basic 10 under Visual Studio 2010 Professional Edition, tested and ran on a MS-Windows XP and 7 workstations. It can be tested on Fedora Linux 20 system using WINE – WINdows Emulator

– from www.winehq.org, to run the executables. The source code files can be viewed in any text editor if the user is using an OS other than Windows (or using Windows without VS installed), but to manipulate and rebuild the programs the user will need other workarounds, like using MonoDevelop from www.go-mono.com/mono-downloads/download.html, for example, as an IDE that can run and compile .NET framework programs). That being said, the programs should run on any system that supports EXE file formats and a .NET framework JIT compiler (Just-In-Time compiler), or a similar software to run the CIL (Common-Intermediate Language) executables.

- At least basic programming knowledge with Visual Basic, although the example programs are straightforward and the language structure is clean enough to make it easy for programmers experienced with other programming languages to follow program logic and run/test the programs.
- Microsoft Visual Studio 2010 Enterprise Edition was used to program, compile, and debug the examples presented in this book. It can be obtained from www.visualstudio.com/en-us/products/visual-studio-express-vs.aspx (for the express edition, a lightweight edition available for the programming languages separately), or from go.microsoft.com/fwlink/?linkid=240162, or getting the software CD/DVD from your software vendor, or by simply looking for "Microsoft Visual Studio Download" in Google search engine.

Being written as GUI (Graphical User Interface) programs for windowing systems, the programs have two parts: the GUI design (what the user sees), and the code doing the work behind the scenes. The code for each program is included in the text along each example presented. The code provided goes into the main program window source file (usually named *Form1.vb* by default, if not indicated otherwise in the text), and as all the programs here are single-windowed, this will be the case. As for the GUI design part, all the user interfaces are included as snapshots in Appendix (3) in

17

the back of the book. The user can refer to them when reading/testing the programs to link the controls used in the design with the code (as the default names assigned to the controls are retained, that is, TextBox1, Label2, Form1, and so on, are left unchanged, to make it easy for the reader to follow what-goes-where).

The programming code was designed to be at the minimum level needed to perform the tasks required, at the same time not to be too short, scrambled, or ambiguous, so that new and novice programmers can follow with ease. For each task performed, especially if a function performs a lengthy operation, or a specific programming concept is complex or advanced, the reader will find prompt documentation through the source code in the relevant parts to ease following the code and make it a rewarding and productive task.

1.3 Visual Basic.NET programming language
Visual Basic.NET was selected as a programming language for the book's examples for many reasons, to mention but a few:

- MS-Windows is one of the most (if not *the* most) popular operating system in use today. Chances are, when buying a new PC or laptop (or even a mobile phone, tablet or iPad), the user will face this OS first and foremost.
- .NET Fx (pronounced "dot net framework") is becoming so popular as a dominant programming platform in computer industry, that it will be counterproductive not to learn and understand how to implement engineering models using this platform.
- Visual Basic.NET came from the old good beginners' friend BASIC, which was infamous for being an easy to learn language, albeit with moderate capabilities, but above all, having so many layers of abstraction that the programmer doesn't need to bother about the low-level workings of the computer, he/she just focuses on programming. For this, BASIC, and its descendent VB, which then evolved to

18

VB.NET, is considered one of the easiest programming languages to learn, especially for beginners.

Visual Basic.NET is not a for-all-purposes language. Praise it enough for good abstraction, clear structure, and educational appeal, the language is a property of Microsoft Corporation. Being a proprietary software, along with the .NET framework itself, it is not easy (not impossible though) to directly use the language to program native software under other operating systems (for example Linux, which is becoming an important player in the arena of operating systems nowadays). Some workarounds includes using OS emulators, or .NET-compatible IDEs and compilers (discussed above). Sometimes it is just as easy to use another programming language like Java or Qt/C++ that supports cross-plat forming out-of-the-box, pros and cons for this scheme are discussed above.

As much as the authors support (and in fact, use) cross-platform model design and encourage young and beginner programmers to explore it. Nevertheless, it was felt that following such a design in this book will add so much overhead to the discussion that it will stray the reader out of the main point of the day: basic model design. That is, the basic function of the program examples in the book is to explain model designs in the most clear and straightforward way, not to show the reader how to become a professional VB programmer. To that end, programmers in other languages are encouraged to translate the example programs to their language of expertise, and to cross-reference with the code example herein, it ought to be a productive and educational task. You are free to use the code and manipulate it as freedom is, as long as it is not used in a commercial-level program, in which case the programmer is kindly asked to reference the source text in their programs.

This technical background is not to undermine the wealth of engineering knowledge invested in this book, that is, the book can (and ought to) be used as an engineering textbook on all its entirety. So skipping the programming algorithms and listings associated with the examples, if the reader doesn't want, or doesn't have the time or tools, to read and test the computer programs presented, is a perfectly plausible and productive way to read the book.

We hope this book will be as informing and appealing to the reader as was intended to, or more.

1.4 Programs on the Accompanying CD/DVD

The programs on the CD accompanying this book include both source code and executable files. They are grouped in folders named by the chapter. A separate folder including the executables is named "EXE Alone", if you want to run the examples without reviewing the source code. To run the CD, simply insert it in the CD drive and it will auto-run. If it did not, open the CD in MyComputer and run SOLID.exe, which is a self-extracting archive. It will uncompress the folders mentioned above.

There should be no problem running the CD and example programs, but for the sake of completion, consider the following system hardware and software specifications as required:

- A PC or a laptop.
- A CD/ROM drive.
- A hard disk drive with at least 5 MB free space (for the code and EXE files).
- A math co-processor chip is recommended to enhance performance (usually incorporated into modern Intel processors).
- A Windows operating system version XP, 2003, 7.
- An installed .NET framework package version 4.x or later (see the User's Manual on how to download and install the latest version).

For users using other operating systems, some solutions were described earlier (using emulators, etc…).

Chapter Two

Solid Waste Classification and Quantities

2.1 Background
Solid waste may be defined as a heterogeneous mass of throwaways from the urban community, as well as the more homogeneous accumulations of agricultural, industrial and mineral wastes. Otherwise it may be defined as materials that do not represent major outputs of market product, and serve no productive or consumptive purpose to its producer, which justifies its disposal. Generally, a solid waste represents those things which are unwanted, useless and not needed by someone.

2.2 Sources of solid waste
There are many sources of solid waste, garbage and sweeping which include: agriculture, mining, building and construction, industry, housing, homes, offices, open markets, restaurants, hospitals, shops, educational institutions ... etc. Solid waste may be classified as:
- Municipal solid waste from a community, MSW, having mixed household and residential waste, garbage, food waste and recyclables (such as: newspapers, aluminum cans, milk cartons, plastic soft drink bottles, steel cans, corrugated cardboard, other material collected by the community).
- Household hazardous waste.
- Commercial waste which may contain many of the same items as generated by household waste. It can arise from design health facilities, hotels, maintenance workshops and warehouses, markets, offices, stores, shops, printing institutions, rest houses, restaurants … etc.
- Yard (or green) waste originating with individual households and animal waste.
- Street refuse, litter and waste from community trash cans produced by individuals, cleanliness of streets, streets, parks and playgrounds sweeping, thrown on the side of the road

21

from users (sweeping), municipal containers, debris and rubble, dead animals (small animals such as: cats and dogs, and big animals such as: horses, sheep, donkeys and cattle), damaged cars placed on both sides of the road.

- Bulky refuse items and white waste (Such as: bicycles, furniture, old and used cars, damaged vehicles, refrigerators and gas and electric stoves, rugs, etc.).
- Construction and demolition waste and ruins of buildings..
- Industrial refuse from places of construction, manufacturing, Sources of metals production and processing, mining, refineries, chemical plants and power plants. It may contain bricks, concrete, dust, stones, mortar, outputs of: refrigeration and air conditioning, plumbing, electricity, water and phone ... etc.
- Solid waste from waste treatment, purification plants, industrial wastewater treatment processes and air pollution control plants.
- Hazardous waste from radioactive materials, chemical, biomaterial, hospitals and medical waste, pathological and infectious, remains of experimental animals, corpses, remnants of drugs, poisons and chemicals and containers
- Agriculture waste, field crops and different farm types from: planting, harvesting of fields, livestock farms producing dairy, meat, slaughterhouses ... etc.
- Other products of the information revolution from damaged computers, peripherals and software CDs, CD-plastic ... etc..

Table (2.1) gives examples of main sources and types of production units of solid waste and garbage.

Table (2.1): Sources and types of production units of solid waste [2,4,6,10,24,25,46].

Source of waste	Production units	Type of solid waste
Residential/ Domestic/municipal	Single-family, multi-family dwellings, houses, low-medium-, and high rise apartments, villa, and housing.	Food waste, rubbish, garbage, ashes, special waste.
Commercial	Stores, shops, restaurants, markets, office buildings, motels, print shops, auto repair shops, medical facilities, …etc.	Food waste, rubbish, ashes, demolition and construction waste, special waste, hazardous waste.
Industrial	Construction, fabrication, light and heavy manufacturing, oil refineries, chemical plants, mining, logging, power plants, demolition, …etc.	Food waste, rubbish, ashes, demolition and construction wastes, special waste, hazardous waste.
Square and open areas	Roads, streets, alleys, parks, playgrounds, beaches, bathing and recreational areas, squares, highways, gardens	Special waste, rubbish.
Agricultural	Field and row crops, fruit orchards vineyards squares, diaries, butter and cheese, laboratories experimental fields, feedlots, farms, … etc.	Spoiled food wastes, agricultural waste, rubbish, hazardous materials.
Treatment plant sites	Water, wastewater, industrial treatment processes, … etc.	Treatment plant wastes, principally composed of residual sludges.

2.3 Amount of municipal solid waste

Equations 2.1 and 2.2 give a rough estimate for amount of municipal solid waste or refuse generated from a community.

MSW, municipal solid waste =
  (refuse) + construction and demolition waste + leaves +
  bulky items   (2.1)
or,
(MSW) = (refuse) + (C and D waste) + (leaves) + (bulky items)
                                                               (2.2)

Refuse can be defined in terms of as-generated and as-collected solid waste. The refuse generated includes all of the wastes produced by a household. Often some part of the refuse, especially organic matter and yard waste, is composted on premises. The fraction of refuse that is generated but not collected is called diverted refuse. The as-generated refuse is always larger than the as-collected refuse, and the difference is the diverted refuse, see equation 2.3.

(As-generated refuse) = (As-collected refuse) + (Diverted refuse)
                                                               (2.3)

Example (2.1)

a) A certain community produces the following quantities of solid waste on an annual basis:

Fraction	Tons per year
Mixed house waste	250
Recyclables	30
Commercial waste	50
Construction and demolition debris	135
Leaves and miscellaneous	40

Generated recyclables are collected separately and processed at a materials recovery plant. Both mixed household and commercial wastes are taken to the municipality landfill, as do the leaves and

24

miscellaneous solid wastes. The C and D wastes are used to fill a large ravine. Calculate percentage of diversion.

b) Write a computer program to calculate the percentage of diversion for a certain community given the quantities of solid waste produced on an annual basis.

c) Verify your program by solving example 2.1.

Solution

1) Given: MSW annual quantities in tons.

2) If the calculation is on the basis of MSW, the total waste generated is 505 tons per year. If everything not going to the landfill is counted as having been diverted, the diversion is calculated as

$$\text{Diversion} = \frac{(30+135+40)}{505} \times 100 = 41\%$$

3) This is an impressive diversion. But if the diversion is calculated as that fraction of the refuse (mixed household and commercial waste) that has been kept out of the landfill by the recycling program, the diversion is

$$\text{Diversion} = \frac{(30)}{250+50} \times 100 = 10\%$$

4) This is not nearly as impressive, but a great deal more honest.

Program 2.1 Algorithm: solid waste diversion to landfill

1. **Inputs:** MSW annual quantities (tons)
2. **Calculations:** See Equation (1.3)
3. **Output:** Percentage of diversion

Program 2.1 Listing:

```
'************************
'EXAMPLE 2.1
'************************
Public Class Form1

    Private Sub Form1_Load(ByVal sender As System.Object,
                    ByVal e As System.EventArgs)
```

```
                                     Handles MyBase.Load
        REM ** set main window's title **
        Me.Text = "Example 1.1"
        Me.FormBorderStyle =
             Windows.Forms.FormBorderStyle.FixedSingle
        REM ** don't enable form size changes **
        Me.MaximizeBox = False
        REM ** prepare the user interface **
        Label1.Text = "Mixed house waste"
        Label2.Text = "Recyclables"
        Label3.Text = "Commercial waste"
        Label4.Text = "Construction and demolition debris"
        Label5.Text = "Leaves and miscellaneous"
        Label6.Text = "Diversion (% of MSW)"
        Label7.Text = "Diversion (% of refuse)"
        Button1.Text = "&Calculate"
        TextBox1.Focus()
    End Sub

    Private Sub Button1_Click(ByVal sender As System.Object,
                        ByVal e As System.EventArgs)
                        Handles Button1.Click
        REM ** define variables used in calculations **
        Dim mixed, rec, comm, construct, misc As Double
        Dim MSW, refuse, diversion1, diversion2 As Double
        REM ** get values of the variables from user input **
        mixed = Val(TextBox1.Text)
        rec = Val(TextBox2.Text)
        comm = Val(TextBox3.Text)
        construct = Val(TextBox4.Text)
        misc = Val(TextBox5.Text)
        REM ** calculate output **
        MSW = mixed + rec + comm + construct + misc
        diversion1 = ((rec + construct + misc) / MSW) * 100
        refuse = mixed + comm
        diversion2 = (rec / refuse) * 100
        REM ** .. and show the output to user
        REM ** formatted with two digits to the right
        REM ** of the decimal point
        TextBox6.Text = FormatNumber(diversion1, 2)
        TextBox7.Text = FormatNumber(diversion2, 2)
    End Sub
End Class
```

26

Exercise (2.1)

1) Select a short research project on solid waste for a certain locality (municipal, industrial, commercial, agricultural, hazardous ... etc.). Write briefly about the following:
   - Selection of research topic and justification.
   - Research objectives, hypothesis and assumptions
   - Research methodology
   - Materials and methods for selected research area with emphasis on: SW sources and characteristics, collection, segregation, sorting, treatment, final disposal and reuse and recycling.
   - Results and discussions
   - Conclusions and recommendations
   - References.

2) Write briefly about any **THREE** of the following: (B.Sc., UoD, 2013)
   a) Problems of waste and garbage.
   b) Sources of waste, garbage & sweeping.
   c) Important characteristics and properties of municipal solid waste. Elaborate on benefits to be gained through their determination.
   d) Factors affecting cost of MSW collection.
   e) Factors affecting effectiveness of magnets to sort ferrous materials from rest of solid waste.

Exercise (2.2)

1) A community produces the following on an annual basis:

Fraction	Tons per year
Mixed house waste	230
Recyclables	25
Commercial waste	45
Construction and demolition debris, C&D	120
Leaves and miscellaneous	50
Treatment plant sludges	5

The recyclables are collected separately and processed at a materials recovery facility. The mixed household waste and the commercial waste go to the landfill, as do the leaves and miscellaneous solid wastes. The sludges are dried and applied on land (not into the landfill), and the C & D wastes are used to fill a large ravine. Calculate the diversion.

2) A community produces the following on an annual basis:

Fraction	Tons per year
Mixed household waste	185
Recyclables	41
Commercial waste	52
Construction and demolition debris	98
Treatment plant sludges	25
Leaves and miscellaneous	5

The recyclables are collected separately and processed at a materials recovery facility. The mixed household waste and the commercial waste go to the landfill, as do the leaves & miscellaneous solid wastes. The sludges are dried & applied on land (not into the landfill), and the C & D wastes are used to fill a large ravine. Calculate the diversion. (B.Sc., UoD, 2013) (Ans. 42, 15%)

Chapter Three

Municipal Solid Waste Properties

3.1 General

Properties of solid waste affect design of solid waste collection systems, treatment and disposal, operation, management and performance of units. Valuable MSW properties may constitute: physical, chemical and biological properties. Physical and material properties of solid waste affect design of storage equipment, its transfer, transpiration and treatment. They may include: grain size, material components, use of material, degree of purity, contents of solid waste, moisture content, grain size calorific value, density, mechanical properties, degree of decomposition … etc.. Chemical properties may include: chemical composition, chemical content … etc.

3.2 MSW properties

Benefits of MSW properties may be summarized as follows:

- Quantifying amounts of waste and hazardous materials generated when considering disposal by landfilling.
- Assessment of useful organic materials for production of beneficial gas.
- Estimating amount of energy when recycling or if materials or energy recovery by combustion is the objective.
- Knowledge of hazardous and harmful substances that may be present in solid waste for its sorting and disposal.
- Knowledge of useful material for incineration and energy access.

Properties of MSW of significance and interest include the following:

1) Physical properties:
    a. Composition by identifiable items (steel cans, office paper, etc.).

29

    b. Weight.
    c. Moisture content.
    d. Particle size and grain size distribution.
    e. Heat and calorific value.
    f. Density.
    g. Angle of stability.
    h. Mechanical properties to evaluate alternative processes and options for energy recovery by focusing on: pressure stress, stress-strain curve for some materials and modulus of elasticity.

2) Chemical properties:
    a. Chemical composition: carbon, hydrogen, concentration of metals.
    b. Proximate analysis.
    c. Fusing point of ash.
    d. Ultimate analysis (major elements).
    e. Compositional analysis.
    f. Calorimetry.
    g. Energy content.
    h. Volatile solids lost upon ignition.
    i. pH value.
    j. Toxic elements.
    k. Nutrients (carbon, nitrogen and phosphorus).

3) Biological properties (biodegradability).

3.3 Physical properties of solid waste
3.3.1 Moisture content

The moisture content becomes important when the refuse is processed into fuel or when it is fired directly. Moisture content influences many MSW properties of importance. The extent of this effect depends on the material. When the moisture level exceeds 50%, the high organic fraction can undergo spontaneous combustion if the material is allowed to stand undisturbed.

Moisture content, on wet basis, is found as presented in equation 3.1.

$$M = \frac{W_w - W_d}{W_w} \times 100 \qquad (3.1)$$

Where:
M = Moisture content, percent (on a wet basis), %
W_w = Initial (wet) weight of sample.
W_d = Final (dry) weight of the sample.

Moisture content may be evaluated on a dry weight basis as shown in equation 3.2.

$$M_d = \frac{W_w - W_d}{w_d} \times 100 \qquad\qquad (3.2)$$

Where:
M_d = Moisture content, percent (on a dry basis), %

Table (3.1) gives moisture content of uncompacted refuse components.

Table (3.1): Moisture content of uncompacted refuse components [2,6,10,24,25,46].

component	Moisture content	
	Range	Typical
Residential		
Aluminum cans	2 - 4	3
Cardboard	4 - 8	5
Fines (dirt, etc.)	6 - 12	8
Food waste	50 - 80	70
Glass	1 - 4	2
Grass	40 - 80	60
Leather	8 - 12	10
Non-ferrous Metal	2 - 4	2
Leaves	20 - 40	30
Paper	4 - 10	6
Plastics	1 - 4	2
Ferrous metals	2 - 6	3
Rubber	1 - 4	2
Steel cans	2 - 4	3
Textiles	6 - 15	10
Wood	15 - 40	20
Yard waste	30 - 80	60
Garden trimmings	30 - 80	60
Commercial		
Food waste	50 - 80	70
Mixed organics	10 - 60	25
Mixed	10 - 25	15
Wooden shipping crates and plant scales	10 - 30	30
Construction (mixed)	2 - 15	8
Dirt, ashes, bricks ... Etc.	6 - 12	8
Municipal waste	15 - 40	

Example (3.1)

a) A residential waste has the components presented in the table. Estimate its moisture concentration using the typical values.

Component	%
Tin cans	50
Paper	20
Rubber	20
Food waste	10

b) Write a computer program to estimate moisture concentration of a residential waste given its components using the typical moisture values.
c) Verify your program by solving example 3.1.

Solution

1) Given: Waste components and percent moisture.
2) Assume a wet sample weighing 100 lb. Set up the tabulation below:

Component	Percent	Moisture from table	Dry weight (based on 100 lb.)
Tin cans	50	3	48.5
Paper	20	6	18.8
Rubber	20	2	19.6
Food waste	10	70	9.3
Total	100		

3) The moisture content (wet basis) would then be = (100 − 96.2)/100 = 3.8 %

Program 3.1 Algorithm: moisture concentration of a residential waste

1. **Inputs:** Waste components, percent moisture (%)
2. **Calculations:** See Equation (2.2) and Table (2.1)
3. **Output:** The moisture content (%)

Program 3.1 Listing:

```
'********************************************************
'Program 3.1: Calculates Moisture conc. of solid waste
'using the typical values.
'********************************************************
Public Class Form1
    Dim comp(27) As String
    Dim moist(27) As Integer

    Private Sub Form1_Load(ByVal sender As System.Object,
                        ByVal e As System.EventArgs)
                        Handles MyBase.Load
        Me.Text = "Program 2.1: Moisture concentration of solid
waste"
        Me.FormBorderStyle =
                    Windows.Forms.FormBorderStyle.FixedSingle
        Me.MaximizeBox = False

        Label1.Text = "Select component:"
        Label2.Text = "Percentage:"
        Label3.Text = "Moisture (from Table):"
        Label4.Text = "Dry weight (based on 100lb.):"
        Label5.Text = ""
        Button1.Text = "&Calculate"

        'DATA FROM TABLE 8.2
        comp(0) = "Residential"
        comp(1) = "  Aluminum cans"
        comp(2) = "  Cardboard"
        comp(3) = "  Fines (dirt, etc.)"
        comp(4) = "  Food waste"
        comp(5) = "  Glass"
        comp(6) = "  Grass"
        comp(7) = "  Leather"
        comp(8) = "  Non-ferrous Metal"
        comp(9) = "  Leaves"
        comp(10) = "  Paper"
        comp(11) = "  Plastics"
        comp(12) = "  Ferrous metals"
        comp(13) = "  Rubber"
        comp(14) = "  Steel cans"
        comp(15) = "  Textiles"
        comp(16) = "  Wood"
        comp(17) = "  Yard waste"
        comp(18) = "  Garden trimmings"
```

34

```
comp(19) = "Commercial"
comp(20) = "  Food waste"
comp(21) = "  Mixed organics"
comp(22) = "  Mixed"
comp(23) = "  Wooden shipping crates and plant scales"
comp(24) = "  Construction (mixed)"
comp(25) = "  Dirt, ashes, bricks ... etc."
comp(26) = "  Municipal waste"
comp(27) = "Select component"

'MOISTURE CONTENT FROM TABLE 8.2
moist(0) = 0
moist(1) = 3
moist(2) = 5
moist(3) = 8
moist(4) = 70
moist(5) = 2
moist(6) = 60
moist(7) = 10
moist(8) = 2
moist(9) = 30
moist(10) = 6
moist(11) = 2
moist(12) = 3
moist(13) = 2
moist(14) = 3
moist(15) = 10
moist(16) = 20
moist(17) = 60
moist(18) = 60
moist(19) = 0
moist(20) = 70
moist(21) = 25
moist(22) = 15
moist(23) = 30
moist(24) = 8
moist(25) = 8
moist(26) = 0

'ADD THE ITEMS INTO THE COMBOBOXES
ComboBox1.Items.Clear()
ComboBox2.Items.Clear()
ComboBox3.Items.Clear()
ComboBox4.Items.Clear()
ComboBox5.Items.Clear()
ComboBox6.Items.Clear()
```

```vbnet
        ComboBox1.Items.AddRange(comp)
        ComboBox2.Items.AddRange(comp)
        ComboBox3.Items.AddRange(comp)
        ComboBox4.Items.AddRange(comp)
        ComboBox5.Items.AddRange(comp)
        ComboBox6.Items.AddRange(comp)

        'DISABLE THE TEXTBOXES
        TextBox2.Enabled = False
        TextBox3.Enabled = False
        TextBox4.Enabled = False
        TextBox5.Enabled = False
        TextBox7.Enabled = False
        TextBox8.Enabled = False
        TextBox10.Enabled = False
        TextBox11.Enabled = False
        TextBox13.Enabled = False
        TextBox14.Enabled = False
        TextBox16.Enabled = False
        TextBox17.Enabled = False
    End Sub

    Sub calculateResults()
        Dim i, M, Ww, Wd As Double
        Dim totalWW, totalWd, moistC As Double

        totalWd = 0
        totalWW = 0
        'Calculate the dry weights of the components
        i = ComboBox1.SelectedIndex
        If i <> 0 And i <> 19 And i <> 27 Then
            Ww = Val(TextBox1.Text)
            M = Val(TextBox2.Text)
            Wd = Ww - (M * Ww / 100)
            TextBox3.Text = Wd.ToString
            totalWd += Wd : totalWW += Ww
        End If
        i = ComboBox2.SelectedIndex
        If i <> 0 And i <> 19 And i <> 27 Then
            Ww = Val(TextBox6.Text)
            M = Val(TextBox5.Text)
            Wd = Ww - (M * Ww / 100)
            TextBox4.Text = Wd.ToString
            totalWd += Wd : totalWW += Ww
        End If
```

```
      i = ComboBox3.SelectedIndex
      If i <> 0 And i <> 19 And i <> 27 Then
          Ww = Val(TextBox9.Text)
          M = Val(TextBox8.Text)
          Wd = Ww - (M * Ww / 100)
          TextBox7.Text = Wd.ToString
          totalWd += Wd : totalWW += Ww
      End If
      i = ComboBox4.SelectedIndex
      If i <> 0 And i <> 19 And i <> 27 Then
          Ww = Val(TextBox12.Text)
          M = Val(TextBox11.Text)
          Wd = Ww - (M * Ww / 100)
          TextBox10.Text = Wd.ToString
          totalWd += Wd : totalWW += Ww
      End If
      i = ComboBox5.SelectedIndex
      If i <> 0 And i <> 19 And i <> 27 Then
          Ww = Val(TextBox15.Text)
          M = Val(TextBox14.Text)
          Wd = Ww - (M * Ww / 100)
          TextBox13.Text = Wd.ToString
          totalWd += Wd : totalWW += Ww
      End If
      i = ComboBox6.SelectedIndex
      If i <> 0 And i <> 19 And i <> 27 Then
          Ww = Val(TextBox18.Text)
          M = Val(TextBox17.Text)
          Wd = Ww - (M * Ww / 100)
          TextBox16.Text = Wd.ToString
          totalWd += Wd : totalWW += Ww
      End If

      moistC = ((totalWW - totalWd) / totalWW) * 100
      Label5.Text = "The moisture content (wet basis) = " +
Format(moistC, "n") + "%"
  End Sub


Private Sub ComboBox1_SelectedIndexChanged(ByVal sender
                    As System.Object, ByVal e
                    As System.EventArgs)
                    Handles ComboBox1.
                    SelectedIndexChanged
      'enter the moisture content (From Table 2.1) into
      'the 'moisture' field
      If ComboBox1.SelectedIndex = 0 Or
```

```vbnet
          ComboBox1.SelectedIndex = 19 Or
          ComboBox1.SelectedIndex = 27 Then
             'These are items not to be selected by the user, so
        'clear the textbox and exit sub
             TextBox2.Text = ""
        Else
             TextBox2.Text =
moist(ComboBox1.SelectedIndex).ToString
        End If
    End Sub

    Private Sub ComboBox2_SelectedIndexChanged(ByVal sender
                   As System.Object, ByVal e
                   As System.EventArgs) Handles
                   ComboBox2.SelectedIndexChanged
        'enter the moisture content (From Table 2.1) into
        'the 'moisture' field
        If ComboBox2.SelectedIndex = 0 Or
        ComboBox2.SelectedIndex = 19 Or
        ComboBox2.SelectedIndex = 27 Then
             'These are items not to be selected by the user, so
clear
        'the textbox and exit sub
             TextBox5.Text = ""
        Else
             TextBox5.Text =
moist(ComboBox2.SelectedIndex).ToString
        End If
    End Sub

    Private Sub ComboBox3_SelectedIndexChanged(ByVal sender
                   As System.Object, ByVal e As System.EventArgs)
                   Handles ComboBox3.SelectedIndexChanged
        'enter the moisture content (From Table 2.1) into
        'the 'moisture' field
        If ComboBox3.SelectedIndex = 0 Or
        ComboBox3.SelectedIndex = 19 Or
        ComboBox3.SelectedIndex = 27 Then
             'These are items not to be selected by the user, so
clear
        'the textbox and exit sub
             TextBox8.Text = ""
        Else
             TextBox8.Text =
moist(ComboBox3.SelectedIndex).ToString
        End If
```

```vb
    End Sub

    Private Sub ComboBox4_SelectedIndexChanged(ByVal sender As
            System.Object, ByVal e As System.EventArgs)
            Handles ComboBox4.SelectedIndexChanged
        'enter the moisture content (From Table 2.1) into
        'the 'moisture' field
        If ComboBox4.SelectedIndex = 0 Or
      ComboBox4.SelectedIndex = 19 Or
      ComboBox4.SelectedIndex = 27 Then
            'These are items not to be selected by the user, so
clear
        'the textbox and exit sub
            TextBox11.Text = ""
        Else
            TextBox11.Text =
moist(ComboBox4.SelectedIndex).ToString
        End If
    End Sub

    Private Sub ComboBox5_SelectedIndexChanged(ByVal sender As
        System.Object, ByVal e As System.EventArgs) Handles
        ComboBox5.SelectedIndexChanged
        'enter the moisture content (From Table 2.1) into
        'the 'moisture' field
        If ComboBox5.SelectedIndex = 0 Or
      ComboBox5.SelectedIndex = 19 Or
      ComboBox5.SelectedIndex = 27 Then
            'These are items not to be selected by the user, so
clear
        'the textbox and exit sub
            TextBox14.Text = ""
        Else
            TextBox14.Text =
moist(ComboBox5.SelectedIndex).ToString
        End If
    End Sub

    Private Sub ComboBox6_SelectedIndexChanged(ByVal sender As
        System.Object, ByVal e As System.EventArgs) Handles
        ComboBox6.SelectedIndexChanged
        'enter the moisture content (From Table 2.1) into
        'the 'moisture' field
        If ComboBox6.SelectedIndex = 0 Or
      ComboBox6.SelectedIndex = 19 Or
```

```vbnet
        ComboBox6.SelectedIndex = 27 Then
            'These are items not to be selected by the user, so
clear
        'the textbox and exit sub
            TextBox17.Text = ""
        Else
            TextBox17.Text =
moist(ComboBox6.SelectedIndex).ToString
        End If
    End Sub

    Private Sub Button1_Click(ByVal sender As System.Object,
            ByVal e As System.EventArgs) Handles
Button1.Click
        calculateResults()
    End Sub
End Class
```

3.3.2 Particle size

The most accurate expression of particle-size distribution is graphical. Nonetheless, several mathematical expressions are used. In water engineering, the particle size of filter sand is expressed using the uniformity coefficient, defined as presented in equation 3.3.

$$UC = \frac{D_{60}}{D_{10}} \qquad\qquad (3.3)$$

Where
UC = Uniformity coefficient.
D_{60} = Particle (sieve) size where 60% of the particles are smaller than that size.
D_{10} = Particle (sieve) size where 10% of the particles are smaller than that size.

3.3.3 Permeability of compacted waste

Hydraulic conductivity of compacted wastes governs movement of liquids and gases in a sanitary landfill. Coefficient of permeability may be determined as presented in equation 3.4.

$$K = C_d^2 \frac{\gamma}{\mu} = k \frac{\gamma}{\mu} \qquad\qquad (3.4)$$

Where:
K = Coefficient of permeability.
Cd = Constant or shape factor, dimensionless.
γ = Specific weight of water.
μ = Dynamic viscosity of water.
k = Intrinsic permeability (or specific) = C_d^2 (Typical values for the intrinsic permeability for compacted solid waste in a landfill are in the range between about 10^{-11} and 10^{-12} m^2 in the vertical direction and about 10^{-10} m^2 in the horizontal direction).

2.3.4 Apparent density

Apparent density may be used in estimating amount of solid waste in some cases and to assess requirements of a sanitary landfill cover material. Apparent density of solid waste and garbage varies greatly with exerted pressure, degree of compaction, level of economic development, concentration of produced waste products, geographic location, and season of the year and storage time.

Overall bulk density for a mixture of materials in a container may be estimated by knowing bulk density of each substance separately. For a mixture of two materials A and B, the bulk density of the mixture can be estimated as shown in equation 3.5.

$$\rho_C = \rho_{A+B} = \frac{\rho_A.V_A + \rho_B.V_B}{V_A + V_B} \tag{3.5}$$

Where:

$\rho_c = \rho_{A+B}$ = Bulk density of the mixture of A and B.

ρ_A = Bulk density of material A.

ρ_B = Bulk density of material B.

V_A = Volume of material A.

V_B = Volume of material B.

Bulk density of the mixture of materials can also be estimated by the mass of materials from equation 3.6.

$$\rho_{A+B} = \frac{M_A + M_B}{\dfrac{M_A}{\rho_A} + \dfrac{M_B}{\rho_B}} \tag{3.6}$$

Where:

M = Mass of the material (pounds or tons in the American Standard System or kilograms or tonnes in the SI system)[3].

---

[3]   1 Ton = 2000 lb and 1 Tonne = 1000 kg.

3.3.5 Angle of Repose

The angle of repose is the angle to the horizontal to which the material will stack without sliding. Sand, for example, has an angle of repose of about $35°$, depending on the moisture content. Because of variable density, moisture, and particle size, the angle of repose of shredded refuse can vary from $45°$ to greater than $90°$.

3.3.6 Size of reduction in volume (Reduction volume)

In design and operation when packaging or compacting solid waste in a landfill it is of value computing size of reduction in volume as outlined in equation 3.7.

$$F = \frac{V_c}{V_o} \tag{3.7}$$

Where:

F = Volume of reduction (remaining ratios of original size as a result of compaction).

V_o = Original Size (initial).

V_c = Volume after compaction

Relationship of reduction volume to apparent density can be found from equation 3.8.

$$F = \frac{V_c}{V_o} = \frac{M/\rho_c}{M/\rho_o} = \frac{\rho_c}{\rho_o}$$

$$\tag{3.8}$$

Where:

ρ_o = Original apparent density.

ρ_c = Apparent density after compaction.

Example (3.2)

a) For illustrative purposes only, assume that refuse has the following components and bulk densities:

Component	Percentage (by weight)	Uncompacted bulk density (lb/ft^3)
Miscellaneous paper	50	3.81

| Garden waste | 25 | 4.45 |
| Glass | 25 | 18.45 |

   Assume that the compaction in the landfill is 1200 lb/yd³ (44.4 lb/ft³). Estimate the percent volume reduction achieved during compaction of the waste. Estimate the overall uncompacted bulk density if the miscellaneous paper is removed.
b) Write a computer program to estimate the percent volume reduction and overall uncompacted bulk density achieved during compaction of a certain refuse given its components and bulk densities.
c) Verify your program by solving example 3.2.

Solution
   1) Given: components and bulk densities.
   2) The overall bulk density prior to compaction is.

$$P_{x_1} = \frac{x_1}{x_1 + y_1} \times 100 = \frac{650}{650 + 50} \times 100 = 90$$

   3) The volume reduction achieved during compaction is

$$E_{x,y} = \sqrt{\frac{x_1}{x_o} \frac{y_2}{y_o}} \times 100 = \sqrt{\frac{650}{720} \frac{230}{280}} \times 100 = 86$$

   4) So the required landfill volume is approximately 11% of the volume required without compaction. If the mixed paper is removed, the uncompacted density is

$$P_{(A+B+D)} = \frac{M_A + M_B + M_D}{\dfrac{M_A}{P_A} + \dfrac{M_B}{P_B} + \dfrac{M_D}{P_D}}$$

$$= \frac{25 + 25}{\dfrac{25}{4.45} + \dfrac{25}{18.45}} = 7.18 \ lbl \ ft^3$$

$$F = \frac{P_o}{P_c} = \frac{7.17 \ lbl \ ft^3}{44.4 \ lbl \ ft^3} = 0.16$$

44

Program 3.2 Algorithm: percent volume reduction and overall uncompacted bulk density during compaction of a refuse

1. **Inputs:** Components and bulk densities (lb/ft³)
2. **Calculations:** See Equations (3.6) and (3.8), and Table (3.1)
3. **Output:** The uncompacted bulk density (lb/ft³)

Program 3.2 Listing:

```
'***************************************************************
'Program 3.2: Estimates percent volume reduction achieved
'during compaction of the waste
'***************************************************************
Public Class Form1
    Dim comp() As String
    Dim perc(), dens() As Double

    Private Sub Form1_Load(ByVal sender As System.Object,
            ByVal e As System.EventArgs) Handles MyBase.Load
        Me.Text = "Program 8.2:"
        Me.FormBorderStyle =
            Windows.Forms.FormBorderStyle.FixedSingle
        Me.MaximizeBox = False

        Label1.Text = "Program 3.2: Estimates percent volume
                            reduction achieved "
        Label1.Text += vbCrLf + "during compaction of the
waste"
        Label2.Text = "Enter each component's name, percentage,
                            and bulk density:"
        Label3.Text = "Enter the compaction in the landfill
(lb/ft3):"
        Label4.Text = ""
        Button1.Text = "&Calculate"

        DataGridView1.Columns.Clear()
        DataGridView1.Columns.Add("CCol", "Component")
        DataGridView1.Columns.Add("PCol", "% (by weight)")
        DataGridView1.Columns.Add("DCol", "Bulk
density(lb/ft3)")
        DataGridView1.AutoResizeColumns()

    End Sub
```

```vb
    Sub calculateResults()
        If DataGridView1.RowCount <= 1 Then
            MsgBox("Enter at least one component!", vbOKOnly,
"Error")
            Exit Sub
        End If

        Dim count As Integer = DataGridView1.RowCount - 1
        Dim nom, denom, rho, rhoc As Double

        ReDim comp(count), perc(count), dens(count)
        For i = 0 To count - 1
            comp(i) = DataGridView1.Rows(i).Cells("CCol").Value
            perc(i) =
Val(DataGridView1.Rows(i).Cells("PCol").Value)
            dens(i) =
Val(DataGridView1.Rows(i).Cells("DCol").Value)
        Next

        nom = 0
        denom = 0
        For i = 0 To count - 1
            nom += perc(i)
            denom += perc(i) / dens(i)
        Next
        rho = nom / denom
        Label4.Text = "The overall bulk density prior to
compaction:" +
                            Format(rho, "##.##").ToString _
                            + " lb/ft3"
        rhoc = Val(TextBox1.Text)
        Label4.Text += vbCrLf + "The volume reduction achieved
                            during compaction: " + _
                            Format((rho / rhoc),
"#0.##").ToString

    End Sub

    Private Sub Button1_Click(ByVal sender As System.Object,
            ByVal e As System.EventArgs) Handles
Button1.Click
        calculateResults()
    End Sub
End Class
```

3.3.7 Material Abrasiveness
MSW and refuse consists of different types of abrasive particles and grains such as sand, glass, metals and rocks. Removal of this abrasive material is often necessary prior to some operations (such as pneumatic conveying) can become practical.

3.4 Chemical properties of solid waste
Chemical properties of solid waste are of value in economics of material or energy recovery. Chemical components of solid waste have significant variability and change due to the heterogeneity of solid waste, geographical location and temporal changes. Typically, solid wastes represent a combination of semi-moist combustible and noncombustible materials. When using solid waste as a fuel its chemical properties of significant include: proximate analysis, fusing point of ash, ultimate analysis (major elements) and energy content.

3.4.1 Fusion point of ash
The fusion point of ash may be defined as that temperature at which the ash resulting from the burning of waste will form a solid (clinker) by fusion and agglomeration. Typical fusion temperatures for the formation of clinker from solid waste range from 1100 to 1200°C.

3.4.2 Proximate analysis
Proximate analyses are to determine percentage (fraction) of volatile organic organics and fixed carbon in solid waste and garbage (fuel).

3.4.3 Ultimate analysis
Ultimate analysis uses the chemical makeup of the fuel to approximate its heat value and it depends on elemental composition.

3.4.4 Volatile solids
Volatile solids can be estimated upon ignition at temperature of 550 °C for 4 hours and then cooling in a dryer. Loss in weight represents

volatile organics, which includes disintegrating organic material and non-decombosable material as reflected in equations 3.9 and 3.10.

Loss in weight = Volatile Organics   (3.9)

VO   = D + ND   (3.10)

Where:
VO = Volatile Organics.
D = Disintegrating organic material.
ND = Non-decombsaple material.

3.4.5 Heat value of refuse

Heat value of refuse is of paramount importance in resource recovery. Heat value is expressed as British thermal unit per pound, Btu[4]/lb, of refuse, or kJ/kg in the SI system. Heat value of refuse and other heterogeneous materials may be measured with a calorimeter. A calorimeter is a device in which a sample is combusted and the temperature rise is recorded. Knowing the mass of the sample and the heat generated by the combustion, the Btu/lb is calculated.

The most popular method using ultimate analysis is the DuLong equation, which originally was developed for estimating the heat value of coal.

Energy values of solid waste and garbage can be estimated by using DuLong equation as shown in equation 3.11.

$$\frac{KJ}{kg} = 337\,C + 1428\left(H - \frac{O}{8}\right) + 9\,S$$

$$(3.11)$$

Where:
C = Carbon, (%).
H = Hydrogen, (%).

---

[4]   1 Btu = heat necessary to raise the temperature of 1 lb of water 1°F

O = Oxygen, (%).
S = Sulfur, (%).
The DuLong formula is cumbersome to use in practice, and it does not give acceptable estimates of heat value for materials other than coal. Total energy content may be determined using the modified DuLong formula as presented in equation 3.12.

$$\frac{BTU}{lb} = 145\,C + 610\left(H_2 - \frac{O_2}{8}\right) + 40\,S + 10\,N \qquad (3.12)$$

Where:
Btu/lb^5 = Total energy.
C = Carbon, percent by weight.
H_2 = Hydrogen, percent by weight.
O_2 = Oxygen, percent by weight.
S = Sulfur, percent by weight.
N = Nitrogen, percent by weight.

Table (3.3) is an illustration of ideal data for final analysis of components of a combustible municipal solid waste.
Another equation for estimating the heat value of refuse using ultimate analysis is illustrated in equation 3.13.

Btu/lb = 144 C + 672 H + 6.2 O + 41.4 S - 10.8 N   (3.13)

Where
C, H, O, S, and N = Weight percentages (dry basis) of carbon, hydrogen, oxygen, sulfur, and nitrogen, respectively, in the combustible fraction of the fuel. The sum of all of these percentages has to add to 100%.

Formulas based on compositional analyses are an improvement over formulas based on ultimate analyses. One such formula is indicated in equation 3.14.

---

5   (Btu/lb) x 2.326 = kJ/kg

$$Btu/lb = 49R + 22.5(G+ P) - 3.3W \qquad (3.14)$$

Where
R = Plastics, percent by weight of total MSW, on dry basis.
G = Food waste, percent by weight of total MSW, on dry basis.
P = Paper, percent by weight of total MSW, on dry basis.
W = Water, percent by weight, on dry basis.

Using regression analysis and comparing the results to actual measurements of heat value, an improved form of a compositional model is suggested by equation 3.15.

$$Btu/lb = 1238 + 15.6R + 4.4P + 2.7G - 20.7W \qquad (3.15)$$

Where
R = Plastics, percent by weight, on dry basis.
P = Paper, percent by weight, on dry basis.
G = Food wastes, percent by weight, on dry basis.
W = Water, percent by weight, on dry basis.

Table (3.3): Ideal data for final analysis of components of a combustible municipal solid waste [2,4,6,10,24,25,46].

Component	Percentages by mass (on dry bases)					
	Carbon	Hydrogen	Oxygen	Nitrogen	Sulfur	Ash
Food waste	48.0	6.4	37.6	2.6	0.4	5.0
Paper	43.5	6.0	44.0	0.3	0.2	6.0
Cardboard	44.0	5.9	44.6	0.3	0.2	5.0
Plastics	60.0	7.2	22.8	-	-	10.0
Textiles	55.0	6.6	31.2	4.6	0.15	2.5
Rubber	78.0	10.0	-	2.0	-	1.00

Leather	60.0	8.0	11.6	10.0	0.4	10.0
Garden trimmings	47.8	6.0	38.0	3.4	0.3	4.5
Timber	49.5	6.0	42.7	0.2	0.1	1.5
Mixture of organic materials	48.5	6.5	37.5	2.2	0.3	5.0
Dirt, ash, bricks etc. ...	26.3	3.0	2.0	0.5	0.2	68.0

Example (3.3)

a) Find approximate chemical formula of the organic component of the sample composition of a solid waste as set out in the following table. Use chemical composition obtained to estimate energy content of this solid waste.

Component	Percent by mass
Garden trimmings	10
Food waste	25
Timber	4
Paper	38
Cardboard	13
Rubber	4
Tin cans	6
Total sum	100

b) Write a computer program to predict the approximate chemical formula of the organic component of composition of a solid waste given its composition. Let the program use the obtained chemical composition to determine the energy content of this solid waste.

c) Verify your program by solving example 3.3.

51

Solution

1) Given: sample composition.
2) Form the following table using typical moisture values.

Composition (1)	Percent by Mass (2)	Moisture content (M), % (from table) (3)	Dry mass (M_d), kg (4)
Garden trimmings	10	60	$\dfrac{10 - W_d}{10} = \dfrac{60}{100}$, $Wd = 4$
Food waste	25	75	$\dfrac{25 - W_d}{25} = \dfrac{75}{100}$, $Wd = 6.25$
Timber	4	20	$\dfrac{4 - W_d}{4} = \dfrac{20}{100}$, $Wd = 3.2$
Paper	38	6	$\dfrac{38 - W_d}{38} = \dfrac{6}{100}$, $Wd = 35.72$
Cardboard	13	5	$\dfrac{13 - W_d}{13} = \dfrac{5}{100}$, $Wd = 12.35$
Rubber	4	2	$\dfrac{4 - W_d}{4} = \dfrac{2}{100}$, $Wd = 3.92$
Tin cans	6	3	$\dfrac{6 - W_d}{6} = \dfrac{3}{100}$, $Wd = 5.82$

3) Determine dry mass of mixture without tin cans =
   4+6.25+3.2+35.72+12.35+3.92+5.82 = 65.44
4) Determine Moisture content of mixture = 94 -65.44 =28.56
   %

Table (b)

Component	Mass, kg
Carbon	33.73
Hydrogen	4.52
Oxygen	26.53
Nitrogen	0.61
Sulfur	0.13
Ash	4.08

5) Change moisture content (H_2O) in previous step to hydrogen and oxygen

Hydrogen = (2/18) * 28.56 = 3.17 kg

Thus hydrogen = 4.52 + 3.17 = 7.69 kg

Oxygen = (16/18) * 28.56 = 25.39 kg

Thus oxygen = 26.53 + 25.39=61.2333

6) Repeat summary of rate in table (c) using additions of hydrogen and oxygen as presented in table (d).

Table (d)

Component	Mass, kg
Carbon	33.73
Hydrogen	7.69
Oxygen	51.92
Nitrogen	0.61
Sulfur	0.13
Ash	4.08
Total	98.16

Component	Mass, kg	Percent (by mass)
Carbon	33.73	33.73/98.16= 34.36%
Hydrogen	7.69	7.83%
Oxygen	51.92	52.89%
Nitrogen	0.61	0.62%
Sulfur	0.13	0.13%
Ash	4.08	4.16%
Total	98.16	100

7) Find energy value for waste from Dulong formula

$$\frac{KJ}{kg} = 337\,C + 1428\left(H - \frac{O}{8}\right) + 9\,S$$

$$\frac{KJ}{kg} = 337 \times 34.36 + 1428\left(7.83 - \frac{52.893}{8}\right) + 9 \times 0.13 = 13320.9$$

Element	Mass, kg	Atomic weight	Number of moles
Carbon	33.73	12	2.81
Hydrogen	7.69	1	7.69
Oxygen	51.92	16	3.25
Nitrogen	0.61	14	0.044
Sulfur	0.13	32	0.004

8) Approximate chemical formula with sulfur

Element	Mole normality Sulfur = 1	Mole normality Nitrogen = 1
Carbon	702.5	63.86
Hydrogen	1922.5	174.77
Oxygen	812.5	73.86
Nitrogen	11	1
Sulfur	1	0

Chemical formula with sulfur is: $C_{702.5}H_{1922.5}O_{812.5}N_{11}S$
Chemical formula without sulfur is: $C_{63.86}H_{174.8}O_{73.9}N$

Program 3.3 Algorithm: chemical formula and energy content of a solid waste

1. **Inputs:** Sample composition (given in Table)
2. **Calculations:** Calculate chemical composition then estimate the energy content of solid waste sample (see Table 3.3)
3. **Output:** The approximate chemical formula (with and with sulfur)

Program 3.3 Listing:

```
'********************************************************
'Program 3.3: Calculates chemical composition and estimates
'energy content of solid waste.
'********************************************************
Public Class Form1
    Dim comp(27) As String
    Dim moist(27) As Integer
    Dim PERC(10, 5) As Double

    Private Sub Form1_Load(ByVal sender As System.Object,
            ByVal e As System.EventArgs) Handles MyBase.Load
        Me.Text = "Program 2.3: Chemical composition of solid
waste"
        Me.FormBorderStyle =
                    Windows.Forms.FormBorderStyle.FixedSingle
        Me.MaximizeBox = False

        Label1.Text = "Select component:"
        Label2.Text = "Percentage:"
        Label3.Text = "Moisture (from Table):"
        Label4.Text = "Dry weight (based on 100lb.):"
        Label5.Text = "Component:"
        Label6.Text = "Mass, kg:"
        Label7.Text = "Carbon"
        Label8.Text = "Hydrogen"
        Label9.Text = "Oxygen"
        Label10.Text = "Nitrogen"
        Label11.Text = "Sulfur"
        Label12.Text = "Ash"
        Label13.Text = ""
        Button1.Text = "&Calculate"

        'DATA FROM TABLE 3.2
        comp(0) = "Residential"
        comp(1) = "  Aluminum cans"
        comp(2) = "  Cardboard"
        comp(3) = "  Fines (dirt, etc.)"
        comp(4) = "  Food waste"
        comp(5) = "  Glass"
        comp(6) = "  Grass"
        comp(7) = "  Leather"
        comp(8) = "  Non-ferrous Metal"
        comp(9) = "  Leaves"
        comp(10) = "  Paper"
        comp(11) = "  Plastics"
```

56

```
comp(12) = "  Ferrous metals"
comp(13) = "  Rubber"
comp(14) = "  Steel cans"
comp(15) = "  Textiles"
comp(16) = "  Wood"
comp(17) = "  Yard waste"
comp(18) = "  Garden trimmings"

comp(19) = "Commercial"
comp(20) = "  Food waste"
comp(21) = "  Mixed organics"
comp(22) = "  Mixed"
comp(23) = "  Wooden shipping crates and plant scales"
comp(24) = "  Construction (mixed)"
comp(25) = "  Dirt, ashes, bricks ... etc."
comp(26) = "  Municipal waste"
comp(27) = "Select component"

'MOISTURE CONTENT FROM TABLE 3.2
moist(0) = 0
moist(1) = 3
moist(2) = 5
moist(3) = 8
moist(4) = 70
moist(5) = 2
moist(6) = 60
moist(7) = 10
moist(8) = 2
moist(9) = 30
moist(10) = 6
moist(11) = 2
moist(12) = 3
moist(13) = 2
moist(14) = 3
moist(15) = 10
moist(16) = 20
moist(17) = 60
moist(18) = 60
moist(19) = 0
moist(20) = 70
moist(21) = 25
moist(22) = 15
moist(23) = 30
moist(24) = 8
moist(25) = 8
moist(26) = 0
```

57

```
'ADD THE ITEMS INTO THE COMBOBOXES
ComboBox1.Items.Clear()
ComboBox2.Items.Clear()
ComboBox3.Items.Clear()
ComboBox4.Items.Clear()
ComboBox5.Items.Clear()
ComboBox6.Items.Clear()
ComboBox7.Items.Clear()

ComboBox1.Items.AddRange(comp)
ComboBox2.Items.AddRange(comp)
ComboBox3.Items.AddRange(comp)
ComboBox4.Items.AddRange(comp)
ComboBox5.Items.AddRange(comp)
ComboBox6.Items.AddRange(comp)
ComboBox7.Items.AddRange(comp)

'DISABLE THE TEXTBOXES
TextBox2.Enabled = False
TextBox3.Enabled = False
TextBox4.Enabled = False
TextBox5.Enabled = False
TextBox7.Enabled = False
TextBox8.Enabled = False
TextBox10.Enabled = False
TextBox11.Enabled = False
TextBox13.Enabled = False
TextBox14.Enabled = False
TextBox16.Enabled = False
TextBox17.Enabled = False
TextBox19.Enabled = False
TextBox20.Enabled = False

'PREPARE THE RADIOBUTTONS AND THE LAST COMBOBOX
RadioButton1.Text = "Select values from Table 3.3:"
RadioButton2.Text = "Enter new values:"
RadioButton2.Checked = True
ComboBox8.Items.Clear()
ComboBox8.Items.Add("--Select waste--")
ComboBox8.Items.Add("Food waste")
ComboBox8.Items.Add("Paper")
ComboBox8.Items.Add("Cardboard")
ComboBox8.Items.Add("Plastics")
ComboBox8.Items.Add("Textile")
ComboBox8.Items.Add("Rubber")
```

```
ComboBox8.Items.Add("Leather")
ComboBox8.Items.Add("Garden trimmings")
ComboBox8.Items.Add("Timber")
ComboBox8.Items.Add("Mixture of organic materials")
ComboBox8.Items.Add("Dirt, ash, bricks etc.")

'DATA FROM TABLE 3.3
'PERCENTAGES OF Carbon, Hydrogen, Oxygen, Nitrogen,
'Sulfur, Ash
Dim percStr As String
percStr = "48,6.4,37.6,2.6,0.4,5.0,"
percStr += "43.5,6,44,0.3,0.2,6,"
percStr += "44,5.9,44.6,0.3,0.2,5,"
percStr += "60,7.2,22.8,0,0,10,"
percStr += "55,6.6,31.2,4.6,0.15,2.5,"
percStr += "78,10,0,2,0,10,"
percStr += "60,8,11.6,10,0.4,10,"
percStr += "47.8,6,38,3.4,0.3,4.5,"
percStr += "49.5,6,42.7,0.2,0.1,1.5,"
percStr += "48.5,6.5,37.5,2.2,0.3,5,"
percStr += "26.3,3,2,0.5,0.2,68,"

Dim last As Integer = 0
For i = 0 To 10
    For j = 0 To 5
        PERC(i, j) = Val(percStr.Substring(last,
            percStr.IndexOf(",", last) - last))
        last = percStr.IndexOf(",", last) + 1
    Next
Next
End Sub

Sub calculateResults()
    Dim i, M, Ww, Wd As Double
    Dim totalWW, totalWd, moistC As Double
    Dim cansWd, cansWw As Double

    totalWd = 0
    totalWW = 0
    'Calculate the dry weights of the components
    i = ComboBox1.SelectedIndex
    If i <> 0 And i <> 19 And i <> 27 Then
        Ww = Val(TextBox1.Text)
        M = Val(TextBox2.Text)
        Wd = Ww - (M * Ww / 100)
        If i = 1 Then
```

```
         'if selection is 'Tin Cans', save it to be subtracted
later
                cansWd = Wd
                cansWw = Ww
            End If
            TextBox3.Text = Wd.ToString
            totalWd += Wd : totalWW += Ww
        End If
        i = ComboBox2.SelectedIndex
        If i <> 0 And i <> 19 And i <> 27 Then
            Ww = Val(TextBox6.Text)
            M = Val(TextBox5.Text)
            Wd = Ww - (M * Ww / 100)
            If i = 1 Then
         'if selection is 'Tin Cans', save it to be subtracted
later
                cansWd = Wd
                cansWw = Ww
            End If
            TextBox4.Text = Wd.ToString
            totalWd += Wd : totalWW += Ww
        End If
        i = ComboBox3.SelectedIndex
        If i <> 0 And i <> 19 And i <> 27 Then
            Ww = Val(TextBox9.Text)
            M = Val(TextBox8.Text)
            Wd = Ww - (M * Ww / 100)
            If i = 1 Then
         'if selection is 'Tin Cans', save it to be subtracted
later
                cansWd = Wd
                cansWw = Ww
            End If
            TextBox7.Text = Wd.ToString
            totalWd += Wd : totalWW += Ww
        End If
        i = ComboBox4.SelectedIndex
        If i <> 0 And i <> 19 And i <> 27 Then
            Ww = Val(TextBox12.Text)
            M = Val(TextBox11.Text)
            Wd = Ww - (M * Ww / 100)
            If i = 1 Then
         'if selection is 'Tin Cans', save it to be subtracted
later
                cansWd = Wd
                cansWw = Ww
```

60

```
            End If
            TextBox10.Text = Wd.ToString
            totalWd += Wd : totalWW += Ww
        End If
        i = ComboBox5.SelectedIndex
        If i <> 0 And i <> 19 And i <> 27 Then
            Ww = Val(TextBox15.Text)
            M = Val(TextBox14.Text)
            Wd = Ww - (M * Ww / 100)
            If i = 1 Then
      'if selection is 'Tin Cans', save it to be subtracted
later
                cansWd = Wd
                cansWw = Ww
            End If
            TextBox13.Text = Wd.ToString
            totalWd += Wd : totalWW += Ww
        End If
        i = ComboBox6.SelectedIndex
        If i <> 0 And i <> 19 And i <> 27 Then
            Ww = Val(TextBox18.Text)
            M = Val(TextBox17.Text)
            Wd = Ww - (M * Ww / 100)
            If i = 1 Then
      'if selection is 'Tin Cans', save it to be subtracted
later
                cansWd = Wd
                cansWw = Ww
            End If
            TextBox16.Text = Wd.ToString
            totalWd += Wd : totalWW += Ww
        End If
        i = ComboBox7.SelectedIndex
        If i <> 0 And i <> 19 And i <> 27 Then
            Ww = Val(TextBox21.Text)
            M = Val(TextBox20.Text)
            Wd = Ww - (M * Ww / 100)
            If i = 1 Then
      'if selection is 'Tin Cans', save it to be subtracted
later
                cansWd = Wd
                cansWw = Ww
            End If
            TextBox19.Text = Wd.ToString
            totalWd += Wd : totalWW += Ww
        End If
```

61

```
totalWd -= cansWd
totalWW -= cansWw
moistC = totalWW - totalWd

'FORM TABLE (B) AS IN THE EXAMPLE
Dim total, Carbon, Hydrogen, Oxygen, Nitrogen,
            Sulfur, Ash As Double
Carbon = Val(TextBox22.Text)
Hydrogen = Val(TextBox23.Text)
Oxygen = Val(TextBox24.Text)
Nitrogen = Val(TextBox25.Text)
Sulfur = Val(TextBox26.Text)
Ash = Val(TextBox27.Text)
'Change moisture content (H2O) in previous step to
hydrogen
'and oxygen
Hydrogen += (2 / 18) * moistC
Oxygen += (16 / 18) * moistC
total = Carbon + Hydrogen + Oxygen + Nitrogen
            + Sulfur + Ash
'Calculate the percentages for the gases
Carbon /= total
Hydrogen /= total
Oxygen /= total
Nitrogen /= total
Sulfur /= total
Ash /= total
'Find energy value for waste from Dulong formula
Dim Dulong As Double
Dulong = (337 * Carbon) +
        (1428 * (Hydrogen - (Oxygen / 8))) + (9 * Sulfur)

'Calculate moles for each gas
Dim molC, molH, molO, molN, molS As Double
molC = Carbon / 12       'Moles = Mass/Atomic weight
molH = Hydrogen / 1      'Moles = Mass/Atomic weight
molO = Oxygen / 16       'Moles = Mass/Atomic weight
molN = Nitrogen / 14     'Moles = Mass/Atomic weight
molS = Sulfur / 32       'Moles = Mass/Atomic weight

'Approximate chemical formula with sulfur
'sulfur will be 1 mole = 0.004 * 250..
'so multiply all moles by 250
Dim molC2, molH2, molO2, molN2, molS2 As Double
molC2 = molC * 250 * 100
```

```
        molH2 = molH * 250 * 100
        molO2 = molO * 250 * 100
        molN2 = molN * 250 * 100
        molS2 = molS * 250 * 100
        Label13.Text = "Chemical formula with sulfur is: C"
                    + Format(molC2, "n") + _
                    " H" + Format(molH2, "n") +
                    " O" + Format(molO2, "n") +
                    " N" + Format(molN2, "n") + " S"
        'Approximate chemical formula without sulfur
        'nitrogen will be 1 mole = 0.044 * 22.73..
        'so multiply all moles by 22.73
        molC2 = molC * 22.73 * 100
        molH2 = molH * 22.73 * 100
        molO2 = molO * 22.73 * 100
        molN2 = molN * 22.73 * 100
        molS2 = molS * 22.73 * 100
        Label13.Text += vbCrLf
            + "Chemical formula without sulfur is: C"
            + Format(molC2, "n") + _
            " H" + Format(molH2, "n") +
            " O" + Format(molO2, "n") +
            " N" + Format(molN2, "n")
    End Sub


    Private Sub ComboBox1_SelectedIndexChanged(ByVal sender As
                    System.Object, ByVal e As System.EventArgs)
                    Handles ComboBox1.SelectedIndexChanged
        'enter the moisture content (From Table 2.2) into
        'the 'moisture' field
        If ComboBox1.SelectedIndex = 0 Or
        ComboBox1.SelectedIndex = 19 Or
        ComboBox1.SelectedIndex = 27 Then
            'These are items not to be selected by the user,
        'so clear the textbox and exit sub
            TextBox2.Text = ""
        Else
            TextBox2.Text =
moist(ComboBox1.SelectedIndex).ToString
        End If
    End Sub


    Private Sub ComboBox2_SelectedIndexChanged(ByVal sender As
                    System.Object, ByVal e As System.EventArgs)
                    Handles ComboBox2.SelectedIndexChanged
        'enter the moisture content (From Table 2.2) into
```

```vbnet
        'the 'moisture' field
        If ComboBox2.SelectedIndex = 0 Or
    ComboBox2.SelectedIndex = 19 Or
    ComboBox2.SelectedIndex = 27 Then
            'These are items not to be selected by the user,
        'so clear the textbox and exit sub
            TextBox5.Text = ""
        Else
            TextBox5.Text =
moist(ComboBox2.SelectedIndex).ToString
        End If
    End Sub

    Private Sub ComboBox3_SelectedIndexChanged(ByVal sender As
            System.Object, ByVal e As System.EventArgs)
Handles
            ComboBox3.SelectedIndexChanged
        'enter the moisture content (From Table 2.2) into
        'the 'moisture' field
        If ComboBox3.SelectedIndex = 0 Or
    ComboBox3.SelectedIndex = 19 Or
    ComboBox3.SelectedIndex = 27 Then
            'These are items not to be selected by the user,
        'so clear the textbox and exit sub
            TextBox8.Text = ""
        Else
            TextBox8.Text =
moist(ComboBox3.SelectedIndex).ToString
        End If

    End Sub

    Private Sub ComboBox4_SelectedIndexChanged(ByVal sender As
            System.Object, ByVal e As System.EventArgs)
Handles
            ComboBox4.SelectedIndexChanged
        'enter the moisture content (From Table 2.2) into
        'the 'moisture' field
        If ComboBox4.SelectedIndex = 0 Or
    ComboBox4.SelectedIndex = 19 Or
    ComboBox4.SelectedIndex = 27 Then
            'These are items not to be selected by the user,
        'so clear the textbox and exit sub
            TextBox11.Text = ""
        Else
```

```
        TextBox11.Text =
moist(ComboBox4.SelectedIndex).ToString
        End If
    End Sub


    Private Sub ComboBox5_SelectedIndexChanged(ByVal sender As
            System.Object, ByVal e As System.EventArgs)
Handles
            ComboBox5.SelectedIndexChanged
        'enter the moisture content (From Table 2.2) into
        'the 'moisture' field
        If ComboBox5.SelectedIndex = 0 Or
    ComboBox5.SelectedIndex = 19 Or
    ComboBox5.SelectedIndex = 27 Then
            'These are items not to be selected by the user,
        'so clear the textbox and exit sub
            TextBox14.Text = ""
        Else
            TextBox14.Text =
moist(ComboBox5.SelectedIndex).ToString
        End If
    End Sub


    Private Sub ComboBox6_SelectedIndexChanged(ByVal sender As
            System.Object, ByVal e As System.EventArgs)
Handles
            ComboBox6.SelectedIndexChanged
        'enter the moisture content (From Table 2.2) into
        'the 'moisture' field
        If ComboBox6.SelectedIndex = 0 Or
    ComboBox6.SelectedIndex = 19 Or
    ComboBox6.SelectedIndex = 27 Then
            'These are items not to be selected by the user,
        'so clear the textbox and exit sub
            TextBox17.Text = ""
        Else
            TextBox17.Text =
moist(ComboBox6.SelectedIndex).ToString
        End If
    End Sub


    Private Sub ComboBox7_SelectedIndexChanged(ByVal sender As
            System.Object, ByVal e As System.EventArgs)
Handles
            ComboBox7.SelectedIndexChanged
        'enter the moisture content (From Table 2.2) into
```

```
          'the 'moisture' field
          If ComboBox7.SelectedIndex = 0 Or
       ComboBox7.SelectedIndex = 19 Or
       ComboBox7.SelectedIndex = 27 Then
              'These are items not to be selected by the user,
          'so clear the textbox and exit sub
              TextBox20.Text = ""
          Else
              TextBox20.Text =
moist(ComboBox7.SelectedIndex).ToString
          End If
      End Sub

      Private Sub Button1_Click(ByVal sender As System.Object,
              ByVal e As System.EventArgs) Handles
Button1.Click
          calculateResults()
      End Sub

      Private Sub RadioButton1_CheckedChanged(ByVal sender As
              System.Object, ByVal e As System.EventArgs)
              Handles RadioButton1.CheckedChanged
          ComboBox8.Enabled = True
      End Sub

      Private Sub RadioButton2_CheckedChanged(ByVal sender As
              System.Object, ByVal e As System.EventArgs)
              Handles RadioButton2.CheckedChanged
          ComboBox8.Enabled = False
          TextBox22.Text = ""
          TextBox23.Text = ""
          TextBox24.Text = ""
          TextBox25.Text = ""
          TextBox26.Text = ""
          TextBox27.Text = ""
      End Sub

      Private Sub ComboBox8_SelectedIndexChanged(ByVal sender As
              System.Object, ByVal e As System.EventArgs)
              Handles ComboBox8.SelectedIndexChanged
          Dim i = ComboBox8.SelectedIndex
          If i <= 0 Then Exit Sub
          TextBox22.Text = PERC(i - 1, 0).ToString
          TextBox23.Text = PERC(i - 1, 1).ToString
          TextBox24.Text = PERC(i - 1, 2).ToString
          TextBox25.Text = PERC(i - 1, 3).ToString
```

```
        TextBox26.Text = PERC(i - 1, 4).ToString
        TextBox27.Text = PERC(i - 1, 5).ToString
    End Sub
End Class
```

Example (3.4)

a) A processed refuse-derived fuel has the following composition. Estimate its heat value.

Component	Fraction by weight, dry basis
Paper	0.3
Food waste	0.1
Plastics	0.2
Glass	0.2
Wood	0.1
Cardboard	0.1

b) Write a computer program to calculate the heat value a processed refuse-derived fuel given its composition.

c) Verify your program by solving example 3.4.

Solution

1) Given: Refuse composition.
2) Form the following table.

Component	Fraction by weight, dry basis	Heat value
Paper	0.3	7200
Food waste	0.1	2000
Plastics	0.2	14000
Glass	0.2	60
Wood	0.1	8000
Cardboard	0.1	7000

3) Estimate the heat value based on the typical values in table = 0.3x7200 + 0.1x2000 + 0.2x14000 + 0.2x60 + 0.1x8000 + 0.1x7200 = 6672 Btu/Ib.

Program 3.4 Algorithm: heat value a processed refuse-derived fuel

1. **Inputs:** Refuse composition (fraction weight)
2. **Calculations:** Estimate the heat value using typical values (see Table 3.3)
3. **Output:** The estimate heat value (Btu/Ib).

Program 3.4 Listing:

```
'****************************************************************
'Program 3.4: calculates percentage of heat value
'****************************************************************
Public Class Form1
    Dim comp(6) As String

    Private Sub Form1_Load(ByVal sender As System.Object,
            ByVal e As System.EventArgs) Handles MyBase.Load
        Me.Text = "Program 2.4: calculates percentage of heat
value"
        Me.FormBorderStyle =
            Windows.Forms.FormBorderStyle.FixedSingle
        Me.MaximizeBox = False

        Label1.Text = "Select component:"
        Label2.Text = "Fraction by weight (dry basis):"
```

```
      Label3.Text = "Heat value:"
      Label4.Text = ""
      Button1.Text = "&Calculate"

      'prepare the selection boxes
      comp(0) = "Paper"
      comp(1) = "Food waste"
      comp(2) = "Plastics"
      comp(3) = "Glass"
      comp(4) = "Wood"
      comp(5) = "Cardboard"
      comp(6) = "--Select comp.--"
      ComboBox1.Items.Clear()
      ComboBox1.Items.AddRange(comp)
      ComboBox1.SelectedIndex = 6
      ComboBox2.Items.Clear()
      ComboBox2.Items.AddRange(comp)
      ComboBox2.SelectedIndex = 6
      ComboBox3.Items.Clear()
      ComboBox3.Items.AddRange(comp)
      ComboBox3.SelectedIndex = 6
      ComboBox4.Items.Clear()
      ComboBox4.Items.AddRange(comp)
      ComboBox4.SelectedIndex = 6
      ComboBox5.Items.Clear()
      ComboBox5.Items.AddRange(comp)
      ComboBox5.SelectedIndex = 6
      ComboBox6.Items.Clear()
      ComboBox6.Items.AddRange(comp)
      ComboBox6.SelectedIndex = 6
      TextBox2.Enabled = False
      TextBox3.Enabled = False
      TextBox5.Enabled = False
      TextBox7.Enabled = False
      TextBox9.Enabled = False
      TextBox11.Enabled = False
   End Sub

 Sub calculateResults()
      Dim heatV As Double = 0
      'check each combobox to see if the user selected a
component,
      'if yes, add it to the total.
      If ComboBox1.SelectedIndex <> 6 Then heatV
          += addHeatValue(1)
      If ComboBox2.SelectedIndex <> 6 Then heatV
```

```
                += addHeatValue(2)
        If ComboBox3.SelectedIndex <> 6 Then heatV
                += addHeatValue(3)
        If ComboBox4.SelectedIndex <> 6 Then heatV
                += addHeatValue(4)
        If ComboBox5.SelectedIndex <> 6 Then heatV
                += addHeatValue(5)
        If ComboBox6.SelectedIndex <> 6 Then heatV
                += addHeatValue(6)
        Label4.Text = "Heat value = " + heatV.ToString + "
Btu/Ib."
    End Sub


    Function addHeatValue(ByVal n As Integer) As Double
        Dim fraction, heat As Double

        Select Case n
            'which combobox is it?
            Case 1
                fraction = Val(TextBox1.Text)
                heat = Val(TextBox2.Text)
                Return fraction * heat
            Case 2
                fraction = Val(TextBox4.Text)
                heat = Val(TextBox3.Text)
                Return fraction * heat
            Case 3
                fraction = Val(TextBox6.Text)
                heat = Val(TextBox5.Text)
                Return fraction * heat
            Case 4
                fraction = Val(TextBox8.Text)
                heat = Val(TextBox7.Text)
                Return fraction * heat
            Case 5
                fraction = Val(TextBox10.Text)
                heat = Val(TextBox9.Text)
                Return fraction * heat
            Case 6
                fraction = Val(TextBox12.Text)
                heat = Val(TextBox11.Text)
                Return fraction * heat
            Case Else
                Return 0
        End Select
    End Function
```

```
Private Sub ComboBox1_SelectedIndexChanged(ByVal sender As
        System.Object, ByVal e As System.EventArgs)
        Handles ComboBox1.SelectedIndexChanged
    'change the 'heat value' according to the selected
component
        'from the combobox.
        Select Case ComboBox1.SelectedIndex
            Case 0 : TextBox2.Text = "7200"
            Case 1 : TextBox2.Text = "2000"
            Case 2 : TextBox2.Text = "14000"
            Case 3 : TextBox2.Text = "60"
            Case 4 : TextBox2.Text = "8000"
            Case 5 : TextBox2.Text = "7000"
        End Select
    End Sub

    Private Sub ComboBox2_SelectedIndexChanged(ByVal sender As
        System.Object, ByVal e As System.EventArgs)
        Handles ComboBox2.SelectedIndexChanged
    'change the 'heat value' according to the selected
component
        'from the combobox.
        Select Case ComboBox2.SelectedIndex
            Case 0 : TextBox3.Text = "7200"
            Case 1 : TextBox3.Text = "2000"
            Case 2 : TextBox3.Text = "14000"
            Case 3 : TextBox3.Text = "60"
            Case 4 : TextBox3.Text = "8000"
            Case 5 : TextBox3.Text = "7000"
        End Select
    End Sub

    Private Sub ComboBox3_SelectedIndexChanged(ByVal sender As
        System.Object, ByVal e As System.EventArgs)
        Handles ComboBox3.SelectedIndexChanged
    'change the 'heat value' according to the selected
component
        'from the combobox.
        Select Case ComboBox3.SelectedIndex
            Case 0 : TextBox5.Text = "7200"
            Case 1 : TextBox5.Text = "2000"
            Case 2 : TextBox5.Text = "14000"
            Case 3 : TextBox5.Text = "60"
            Case 4 : TextBox5.Text = "8000"
            Case 5 : TextBox5.Text = "7000"
```

```
            End Select
      End Sub

      Private Sub ComboBox4_SelectedIndexChanged(ByVal sender As
            System.Object, ByVal e As System.EventArgs)
            Handles ComboBox4.SelectedIndexChanged
      'change the 'heat value' according to the selected
component
            'from the combobox.
            Select Case ComboBox4.SelectedIndex
                Case 0 : TextBox7.Text = "7200"
                Case 1 : TextBox7.Text = "2000"
                Case 2 : TextBox7.Text = "14000"
                Case 3 : TextBox7.Text = "60"
                Case 4 : TextBox7.Text = "8000"
                Case 5 : TextBox7.Text = "7000"
            End Select
      End Sub

      Private Sub ComboBox5_SelectedIndexChanged(ByVal sender As
            System.Object, ByVal e As System.EventArgs)
            Handles ComboBox5.SelectedIndexChanged
      'change the 'heat value' according to the selected
component
            'from the combobox.
            Select Case ComboBox5.SelectedIndex
                Case 0 : TextBox9.Text = "7200"
                Case 1 : TextBox9.Text = "2000"
                Case 2 : TextBox9.Text = "14000"
                Case 3 : TextBox9.Text = "60"
                Case 4 : TextBox9.Text = "8000"
                Case 5 : TextBox9.Text = "7000"
            End Select
      End Sub

      Private Sub ComboBox6_SelectedIndexChanged(ByVal sender As
            System.Object, ByVal e As System.EventArgs)
            Handles ComboBox6.SelectedIndexChanged
      'change the 'heat value' according to the selected
component
            'from the combobox.
            Select Case ComboBox6.SelectedIndex
                Case 0 : TextBox11.Text = "7200"
                Case 1 : TextBox11.Text = "2000"
                Case 2 : TextBox11.Text = "14000"
                Case 3 : TextBox11.Text = "60"
```

```
            Case 4 : TextBox11.Text = "8000"
            Case 5 : TextBox11.Text = "7000"
        End Select
    End Sub

    Private Sub Button1_Click(ByVal sender As System.Object,
            ByVal e As System.EventArgs) Handles
Button1.Click
        calculateResults()
    End Sub
End Class
```

Exercise (3.1)

1) What is the benefit of properties of solid waste in management systems and related engineering topics?
2) How can you estimate the amount of solid waste in an area?
3) What are the related effects to physical properties of solid waste and garbage?
4) What is the benefit to know the angle of stability in a landfill?
5) Write briefly about **THREE** of the following: (B.Sc., UoD, 2012)
   a. Current classification of solid waste in KSA.
   b. The challenge for society is to minimize how much waste is generated and to convert waste into a resource (Essence of the zero waste concept).
   c. Potential problems of solid waste, garbage and sweeping.
   d. Factors that affect quality and quantity of solid waste produced from a particular locality.
   e. Most important properties of solid waste and their significance.
6) Attempt writing briefly about **ANY THREE** questions of the following: (B.Sc., UoD, 2012)
   i.   "The challenge for society is to minimize how much waste is generated and to convert waste into a resource". Discuss this statement.
   ii. Municipal solid waste may be defined as a "heterogeneous mass of throwaways from the urban community, as well as the more homogeneous accumulations of agricultural, industrial and mineral wastes". Based on this definition, how can you classify urban municipal solid waste? State your reasons.
   iii. "Waste and garbage disposal is a big responsibility for the government. If the authority did not have good management for its disposal, it exposes itself to political and social problems." Explain why?
   iv. "It is difficult to determine relevance of diseases with waste and garbage. Nonetheless, about 50% of various diseases are transferred by flies, mosquitoes and rodents proliferating in

the waste." To take caution, what procedures would you advocate to be followed by concerned authorities?

v.   There are many sources of solid waste, garbage and sweeping which include: agriculture, mining, building and construction, industry, housing, homes, offices, open markets, restaurants, hospitals, shops, educational institutions, hazardous ... etc. Outline major types of hazardous solid waste. Which type would expect to be found in KSA? Why?

vi.   Write briefly about most important properties of solid waste and associated benefits.

7) Indicate importance of moisture content measurements for a sample of municipal solid waste. (B.Sc., UoD, 2012)

8) List a method of conducting an experiment to estimate the moisture content of a household solid waste.

9) Why do newspapers contain higher moisture content compared to plastic materials in a domestic dustbin?

10) What is the benefit to know the size of grains of a solid waste?

11) How bulk density of trade solid waste is determined?

12) How can you estimate the chemical composition of garbage?

13) What is the benefit of estimating the calorific value of a solid waste?

14) What is the purpose of measuring heat values of refuse? (B.Sc., UoD, 2012)

15) Write briefly about **THREE** of the following: (B.Sc., UoD, 2013)

a) Classification of solid waste in KSA. Comment on this categorization.

b) Problems and malfunctions of solid waste in your locality. Give suggestions for their reduction.

c) Factors that affect quality and quantity of solid waste produced within your area of residence. Offer appropriate recommendations for improvement.

d) Most significant properties of solid waste. State reasons for their importance.

Exercise (3.2)

1) A residential waste has the following components:

Aluminum cans	10%
Paper	35%
Glass	15%
Food	30%
Plastic	10%

Estimate its moisture concentration using the typical values in table of moisture content.

2) Household garbage contains the following components

Tin cans	10%
Paper	30%
Leather	10%
Food waste	30%
Cardboard	20%

Estimate moisture content using the typical values in a table.

3) A residential waste has the components presented in the table. Estimate its moisture concentration using the typical values. (B.Sc., UoD, 2012)

Paper	40%
Steel cans	10%
Food	30%
Garden trimmings	10%
Leather	10%

4) A residential waste has the following percent composition:

Aluminum cans	10
Paper	40
Glass	20
Food	20
Plastic	10

Assume a wet sample weighing 100 weight unit. For waste moisture content (wet basis) of 17.3 % determine ideal moisture concentration of food component, using the typical values in table of moisture content. (B.Sc., UoD, 2013) (Ans. 70%)

5) Estimate the overall moisture content of a sample of as collected residential municipal solid waste with the typical composition given in table (1). (B.Sc., UoD, 2013) (Ans. 22%)

Typical physical composition of residential municipal solid waste

	Percent by weight	
Component	Range	Typical
Organic		
Food wastes	6 – 18	11
Paper	25 – 40	30
Cardboard	3 – 10	7
Plastics	4 – 10	8
Textiles	0 – 4	2
Rubber	0 – 2	1
Leather	0 – 2	1
Yard wastes	5 – 20	17
Wood	1 – 4	3
Miscellaneous organics	-	-
Inorganic		
Glass	4 – 12	7
Tin cans	2 – 8	7
Aluminum	0 – 1	1
Other metal	1 – 4	2
Dirt, ash, etc.	0 - 6	3
Total		100

6) For illustrative purposes only, assume that refuse has the following components and bulk densities

Component	Percentage(by weight)	Uncompacted bulk density
Miscellaneous paper	50	3.81 (lb/ft^3)
Cardboard	10	1161(lb/yd^3)
Garden waste	20	4.45 (lb/ft^3)
Glass	20	18.45(lb/ft^3)

Assume that the compaction in the landfill is 1300 lb/yd³. (B.Sc., UoD, 2012)

    a) Estimate the percent volume reduction achieved during compaction of the waste using the following equation:

$$P = \frac{2Y_{max}}{\left(\dfrac{q}{k}\right)\left[\dfrac{k.\tan^2 \alpha}{q} + 1 - \dfrac{k.\tan \alpha}{q}\left(\tan^2 \alpha + \dfrac{q}{k}\right)^{\frac{1}{2}}\right]} =$$

¿

$$\frac{2 \times 16}{\left(\dfrac{0.00025}{0.01}\right)\left[\dfrac{0.01 \times (0.02)^2}{0.00025} + 1 - \dfrac{0.01 \times 0.02}{0.00025}\left(t(0.02)^2 + \dfrac{0.00025}{0.01}\right)^{\frac{1}{2}}\right]}$$

    b) Estimate the overall uncompacted bulk density if the miscellaneous paper is removed. Estimate the percent volume reduction achieved.

    c) Comment on your results

7) Assume that a certain refuse has the following components and bulk densities (B.Sc., UoD, 2012)

Component	Percentage (by weight)	Apparent density before compaction (g/cm³)
Aluminum	20	0.038
Glass	10	0.295
Various paper	30	0.061
̶ ̶ ̶ ̶	40	0.268

    i. Assuming compaction of landfill 700 kg/m³, find percent volume reduction achieved when compacting this solid waste.

    ii. Compute total apparent density before compaction by removing various paper component. Estimate the percent volume reduction achieved.

    iii. Comment on your results

8) Estimate the overall moisture content of a sample of as collected residential municipal solid waste with the typical composition given in table (1). (UoD, B.Sc. 2012)

Table (1): Typical physical composition of residential municipal solid waste

Component	Percent by weight	
	Range	Typical
Organic		
Food wastes	6 – 18	9.0
Paper	25 – 40	34.0
Cardboard	3 – 10	6.0
Plastics	4 – 10	7.0
Textiles	0 – 4	2.0
Rubber	0 – 2	0.5
Leather	0 – 2	0.5
Yard wastes	5 – 20	18.5
Wood	1 – 4	2.0
Miscellaneous organics	-	-
Inorganic		
Glass	4 – 12	8.0
Tin cans	2 – 8	6.0
Aluminum	0 – 1	0.5
Other metal	1 – 4	3.0
Dirt, ash, etc.	0 – 6	3.0
Total		100

9) A certain solid waste have the following composition and apparent density:

Component	Percentage (by weight)	Apparent density before compaction (gm/cm³)
Aluminum	10	0.038
Glass	10	0.295
Various paper	40	0.061
Food waste	40	0.368

Assuming compaction of landfill 700 kg/m³, find percent volume reduction achieved when compacting this solid waste. Compute total apparent density before compaction by removing various paper components.

10) The following table shows the components and bulk density of a certain solid waste and garbage.

Component	Percent (by weight)	Apparent density before compaction, g/cm³
Yard waste	20	0.071
Plastics	10	0.037
Newspaper	20	0.099
Glass	10	0.295
Food waste	30	0.368
Corrugated cardboard	10	0.03

Assuming compaction in a landfill for the production of apparent density in the field of 700 kg/m³, find size reduction due to compaction of the solid waste. Find apparent density before compaction assuming total separation of glass and newspapers.

11) A certain solid waste have the following composition & apparent density:

Component	Percentage (by weight)	Apparent density before compaction (gm/cm³)
Aluminum	15	0.038
Food waste	35	?
Glass	15	0.295
Various paper	35	0.061

Determine apparent density of food waste before compaction. Assume compaction of landfill is 700 kg/m³ and 12.8 % volume reduction achieved when compacting this solid waste (i.e. volume of landfill needed is 12.8 % of volume needed without compaction), (B.Sc., UoD, 2013) (Ans. 0.368 g/cm³)

12) Find approximate chemical formula of the organic component of the sample composition of a solid waste as set out in the

following table. Use chemical composition obtained to estimate energy content of this solid waste.

Component	Percent by mass
Garden trimmings	10
Food waste	20
Timber	5
Paper	35
Cardboard	15
Rubber	10
Tin cans	5
Total sum	100

Assume total organic composition of the solid waste assuming a mass of 100 kg of the sample as shown in table (b).

Component	Mass, kg
Carbon	35
Hydrogen	5
Oxygen	28
Nitrogen	0.5
Sulfur	0.1
Ash	5

13) Find the approximate chemical formula for the organic component for a sample of solid waste of the composition set out in the following tables. Use the chemical composition to estimate gross energy content of this solid waste.

Component	Percent by mass
Garden trimmings	10
Food waste	25
Wood	5
Paper	35
cardboard	10
Plastics	10
Glass	5
Grand Total	100

Component	Mass, kg
Carbon	31
Hydrogen	4.5
Oxygen	26
Nitrogen	0.6
Sulfur	0.1
Ash	4

14) Determine the energy value of a typical residential MSW with the average composition shown in the table[6].

Table (1): Typical physical composition of residential municipal solid waste

Component	Percent by weight	
	Range	Typical
Organic		
Food wastes	6 – 18	9.0
Paper	25 – 40	34.0
Cardboard	3 – 10	6.0
Plastics	4 – 10	7.0
Textiles	0 – 4	2.0
Rubber	0 – 2	0.5
Leather	0 – 2	0.5
Yard wastes	5 – 20	18.5
Wood	1 – 4	2.0
Miscellaneous organics	-	-
Inorganic		
Glass	4 – 12	8.0
Tin cans	2 – 8	6.0
Aluminum	0 – 1	0.5
Other metal	1 – 4	3.0
Dirt, ash, etc.	0 - 6	3.0
Total		100

---

[6]   This problem has been adopted from Tichobanoglous, et al (39)..

Table (2): Typical values for energy content of residential municipal solid waste

Component	Energy[7], Btu/lb	
	Range	Typical
Organic		
Food wastes	1500 - 3000	2000
Paper	5000 - 8000	7200
Cardboard	6000 - 7500	7000
Plastics	12000 - 16000	14000
Textiles	6500 - 8000	7500
Rubber	9000 - 12000	10000
Leather	6500 - 8500	7500
Yard wastes	1000 – 8000	2800
Wood	7500 - 8500	8000
Miscellaneous organics	-	-
Inorganic		
Glass	50 - 100[8]	60
Tin cans	100 - 500[(1)]	300
Aluminum	-	-
Other metal	100 - 500[(1)]	300
Dirt, ash, etc.	1000 – 5000	3000
Municipal solid wastes	4000 – 6000	5000

15) Given that the chemical composition of a residential municipal solid waste including sulfur and water is:

$C_{760}H_{1980}O_{874.7}N_{12.7}S$

Determine the total energy content using the modified Dulong formula: (B.Sc., UoD, 2012)

$$\frac{BTU}{lb} = 145\,C + 610\left(H_2 - \frac{O_2}{8}\right) + 40\,S + 10\,N$$

Where:

C = Carbon, percent by weight.
H_2 = Hydrogen, percent by weight.
O_2 = Oxygen, percent by weight.
S = Sulfur, percent by weight.
N = Nitrogen, percent by weight.

---

[7]   As discarded basis.
[8]   Energy content is from coatings, labels, and attached materials.

16) Determine the chemical composition of the organic fraction, without and with sulfur and without and with water, of a residential MSW with the typical composition shown in the table[9].

Typical data on the ultimate analysis of the combustible componentsin residential MSW

Component	Percent by weight (dry basis)					
	Carbon	Hydrogen	Oxygen	Nitrogen	Sulphur	Ash
Organic						
Food wastes	48.0	6.4	37.6	2.6	0.4	5.0
Ppaer	43.5	6.0	44.0	0.3	0.2	6.0
Cardboard	44.0	5.9	44.6	0.3	0.2	5.0
Plastics	60.0	7.2	22.8	-	-	10.0
Textiles	55.0	6.6	31.2	4.6	0.15	2.5
Rubber	78.0	10.0	-	2.0	-	10.0
Leather	60.0	8.0	11.6	10.0	0.4	10.0
Yard wastes	47.8	6.0	38.0	3.4	0.3	4.5
Wood	49.5	6.0	42.7	0.2	0.1	1.5
Inorganic						
Glass	0.5	0.1	0.4	<0.1	-	98.9
Metals	4.5	0.6	4.3	<0.1	-	90.5
Dirt, ash, etc.	26.3	3.0	2.0	0.5	0.2	68.0

17) For a sample of a solid food waste the composition presented in table (a) was detected. Assuming total organic composition of the solid waste and for a mass of 100 kg of the sample as shown in table (b), (B.Sc., UoD, 2012)

- Find moisture content of the mixture.
- Determine approximate chemical formula of the organic component of the sample.
- Use chemical composition obtained to estimate energy content of this waste.

---

[9]   This problem has been adopted from Tichobanoglous, et al (39).

Table (a) Waste composition.

Component	Percent by mass	
Yard waste	10	Table (b)
Food waste	28	Organic
Timber	4	
Paper	40	
Cardboard	10	
Leather	4	
Aluminum cans	4	
Total sum	100	

composition of sample.

Component	Mass, kg
Carbon	30.87
Hydrogen	4.26
Oxygen	21.44
Nitrogen	0.58
Sulfur	0.15
Ash	5.12

Comment on your answers.

18) Define the terms and show how to use the following equations for determining moisture content and energy values of a sample of municipal solid waste.

$$M = \frac{W_w - W_d}{W_w} \times 100$$

$$\frac{KJ}{kg} = 337\,C + 1428\left(H - \frac{O}{8}\right) + 9\,S$$

For a sample of a solid food waste the composition presented in table (a) was detected. Assuming total organic composition of the solid waste and for a mass of 100 kg of the sample as shown in table (b),

- Find moisture content of the mixture.
- Determine approximate chemical formula of the organic component of the sample.

85

- Use chemical composition obtained to estimate energy content of this waste.

Table (a) Waste composition.

Component	Percent by mass
Yard waste	12
Food waste	20
Timber	6
Paper	40
Cardboard	10
Rubber	8
Tin cans	4
Total sum	100

Table (b) Organic composition of sample.

Component	Mass, kg
Carbon	31
Hydrogen	4.3
Oxygen	21.5
Nitrogen	0.5
Sulfur	0.15
Ash	5.1

Comment on your answers.

19) A processed refuse-derived fuel has the following composition.

Component	Fraction by weight, dry basis
Paper	0.3
Food waste	0.3
Plastics	0.2
Glass	0.2

Estimate its heat value.

20) The following table shows a summary of data of total organic composition of a solid waste assuming a mass of 100 kg of the sample. Find molar concentrations and determine approximate chemical formula with sulfur and without sulfur. (B.Sc., UoD, 2013)

Component	Mass, kg
Carbon	33.73
Hydrogen	7.46
Oxygen	50.05
Nitrogen	0.61

(Ans. chemical formula with sulfur is: $C_{702.5}H_{1865}O_{782.5}N_{11}S$.
Chemical formula without sulfur is: $C_{63.9}H_{169.5}O_{71.1}N$)

21) A sample of refuse is analyzed and found to contain 15% water (measured as weight loss on evaporation). The Btu of the entire mixture is measured in a calorimeter and is found to be 5000 Btu/lb. A 1.0 g sample is placed in the calorimeter, and 0.3 g ash remains in the sample cup after combustion. What is the comparable, moisture free Btu and the moisture- and ash-free heat value?

22) A benzoic acid pellet weighing 6.54 g is placed in a bomb calorimeter along with 0.35 g fuse wire. The benzoic acid is ignited, and the temperature rise is 3.6°C. What is the heat capacity of this calorimeter?

23) A 15 g sample of a refuse-derived fuel (RDF) is combusted in a calorimeter that has a heat capacity of 8900 cal/⁰C. The detected temperature rise is 3.5°C. What is the heat value of this sample?

Chapter Four

Solid Waste Collection

4.1 General

Municipal solid waste collection systems are customarily person/truck systems. MSW collectors traverse a generation sites and source production in trucks to transport collected refuse to a site at which the truck is emptied. This process may be an intermediate stopover where the refuse is transferred from a small truck into trailers, larger vans, barges, or railway cars for long-distance transport or a selected final site such as the landfill, compost site, or materials recovery facility. During this cycle some of the useful MSW may be isolated, sorted out or segregated for reuse or recycling or conversion into other useful products.

4.2 MSW collection

The process of refuse collection is a multiphase process, and it can be divided in separate phases namely: House to can (transferring MSW from home to dust pin inside or outside the house), can to truck (for movement of MSW from dust bin to MSW and garbage car by MSW workers or owner of housing), truck from house to house (Collection phase of MSW from different sources by best and efficient ways and its transfer to collection areas and to areas of intermediate or final disposal), truck routing (Stage of path of truck through the city's road network) and truck to disposal (stage of final disposal or recovery of materials).

Example (4.1)

a) A family of six people generates solid waste at a rate of 2.5 lb./cap/day and the bulk density of refuse in a typical garbage can is about 230 lb./yd^3. If collection is once a week, how many 30-gallon garbage cans will they need, or the alternative, how many compacted 20-lb blocks would the family produce if they had a home compactor? How many cans would they need in that case?

b) Write a computer program to calculate number of 30-gallon garbage cans or compacted 20-lb blocks (if they had a home compactor) needed by family given number of household, rate of daily per capita generation of solid waste, bulk density of refuse in a typical garbage can, and frequency of weekly collection.

c) Verify your program by solving example 4.1.

Solution

1) Given: $P = 6$, generated waste = 2.5 lb./cap/day, $\rho = 230$ lb./yd^3.

2) Weight of SW generated = 2.5 lb./cap/day x 6 persons x 7 days/week = 105 Ib.

3) Volume of SW = Weight/density = 105 lb./230 lb./yd^3 = 0.457 yd^3

   Volume (convert to gallons) = 0.457 yd^3 x 202 gal/yd^3 = 92.3 gal

   They will require four 30-gallon cans.

4) If the refuse is compacted into 20-lb blocks, they would need to produce six such compacted blocks to take care of the week's refuse.

   If each block of compacted refuse is 1400 lb./yd^3, the necessary volume is 105 lb./1400 lb./yd^3) x 202 gal/yd^3 = 15.15 gal.

   They would need only one 30-gal can.

Program 4.1 Algorithm: number of 30-gallon garbage cans or compacted 20-lb blocks needed by a household family

1. **Inputs:** Family size (persons), waste generated/day (lb./cap), refuse density in garbage can (lb./yd^3)

2. **Calculations:** Weight of SW, Volume of SW

3. **Output:** Number of needed cans in two cases: (1) with, and (2) without home compactor

Program 4.1 Listing:
```
'****************************************************************
*******************
'Program 4.1:Calculates number of cans and compacted blocks
produced by a family
'****************************************************************
*******************
Public Class Form1

    Private Sub Form1_Load(ByVal sender As System.Object,
            ByVal e As System.EventArgs) Handles MyBase.Load
        Me.Text = "Program 3.1:Calculates number of cans and
            compacted blocks produced by a family "
        Me.FormBorderStyle = BorderStyle.FixedSingle
        Me.MaximizeBox = False

        Label1.Text = "Number of people in the family:"
        Label2.Text = "Rate of daily generation of solid waste
                (lb./capita/day):"
        Label3.Text = "Bulk density of refuse in a typical
garbage can
                (lb/yd3):"
        Label4.Text = "Collection frequency (per week):"
        Label5.Text = ""
        button1.text = "&Calculate"

        Me.Height = 188
    End Sub

    Sub calculateResults()
        Dim s As String = ""
        Dim P, rate, dens, freq As Double
        Dim weight, vol, vol2, vol3 As Double
        P = Val(TextBox1.Text)
        rate = Val(TextBox2.Text)
        dens = Val(TextBox3.Text)
        freq = Val(TextBox4.Text)

        's = "Given: P = " + P.ToString + ", generated waste =
"
                + rate.ToString + " lb./cap/day," _
        '    + " bulk density (rho) = " + dens.ToString + "
lb/yd3."
        weight = rate * P * (7 / freq)   'convert frequency per
                week to days
        s += vbCrLf + "Weight of SW generated =
                90
```

```vb
                    " + rate.ToString + " lb./cap/day x " +
P.ToString + _
                    " persons x " + (7 / freq).ToString + " days/week
= " +
                    weight.ToString + " Lb."
        vol = weight / dens
        s += vbCrLf + "Volume of SW = Weight/density = " +
                    weight.ToString + " lb./" + dens.ToString _
                    + " lb./yd3 = " + vol.ToString + " yd3"
        vol2 = vol * 202   'convert to gallons
        s += vbCrLf + "Volume (convert to gallons) = " +
vol.ToString +
                    " yd3 x 202 gal/yd3 = " _
                    + vol2.ToString + " gal"
        s += vbCrLf + vbCrLf + "They will require " +
                    numberInLetters(Math.Ceiling(vol2 / 30)) +
                    " 30-gallon cans. "
        s += vbCrLf + "If the refuse is compacted into 20-lb
blocks,
                    they would need to "
        s += vbCrLf + "produce such compacted blocks to take
care
                    of the week's refuse"
        vol3 = (weight / 1400) * 202
        s += vbCrLf + vbCrLf + "If each block of compacted
refuse
                    is 1400 lb./yd3, the necessary volume is "
        s += vbCrLf + "(" + weight.ToString
                    + " lb./1400 lb./yd3) x 202 gal/yd3 = " + _
                    Format(vol3, "##.##").ToString + " gal"
        s += vbCrLf + "They would need only " +
            numberInLetters(Math.Ceiling(vol3 / 30)) + " 30-gal
can."

        Label5.Text = s
        Me.Height = 372
    End Sub

    '********************************************************
    '********************************************************
    'takes a number as integer, and returns it in letters.
    'e.g. 1 -> 'One', 2 -> 'Two' and so one.
    '********************************************************
    '********************************************************
    Function numberInLetters(ByVal n As Integer) As String
        Select Case n
```

```
            Case 0 : Return "Zero"
            Case 1 : Return "One"
            Case 2 : Return "Two"
            Case 3 : Return "Three"
            Case 4 : Return "Four"
            Case 5 : Return "Five"
            Case 6 : Return "Six"
            Case 7 : Return "Seven"
            Case 8 : Return "Eight"
            Case 9 : Return "Nine"
            Case 10 : Return "Ten"
            Case 11 : Return "Eleven"
            Case 12 : Return "Twelve"
            Case 13 : Return "Thirteen"
            Case 14 : Return "Fourteen"
            Case 15 : Return "Fifteen"
            Case Else : Return ""
        End Select
    End Function

    Private Sub Button1_Click(ByVal sender As System.Object,
            ByVal e As System.EventArgs) Handles
Button1.Click
        calculateResults()
    End Sub
End Class
```

Number of collection vehicles needed for a community may be
determined from equation 4.1.

$$N = \frac{S * F}{X * W}$$

(4.1)

Where:

N = Number of collection vehicles needed.

S = Total number of customers serviced.

F = Collection frequency, number of collections per week.

X = Number of customers a single truck can service per day.

W = Number of workdays per week.

Example (4.2)

a) Outline how to use the following equation to estimate number of MSW collection vehicles needed for a certain community.

$$N = \frac{SF}{XW}$$

b) Calculate the number of collection vehicles a community would need if it has a total of 4000 services (customers) that are to be collected once per week during working days in a city in KSA. (Realistically, most trucks can service only about 200 to 300 customers before the truck is full and a trip to the landfill is necessary).
Solution

c) Write a computer program to calculate number of collection vehicles a community would need given number of total services (customers) that are to be collected, frequency of weekly collection, and working week days in a certain city, number of customers served by a garbage truck before it is full to make a trip to the landfill.

d) Verify your program by solving example 4.2.

Solution

1) Given:

    N = Number of collection vehicles needed

    S = Total number of customers serviced = 4000

    F = Collection frequency, number of collections per week = 1

    X = Number of customers a single truck can service per day (A single truck can service 300 customers in a single day and still have time to take the full loads to the landfill) = 300.

2) W = Number of workdays per week (The town wants to collect on Saturday, Sunday, Monday, and Tuesdays leaving Wednesdays for special projects and truck maintenance) = 4 days.

3) Thus: N = SF/XW = (4000 * 1)/(300 * 4) = 3.3

4) The community will need four trucks.

Program 4.2 Algorithm: number of collection vehicles needed by a community

1. **Inputs:** Customers' number (S), collection frequency (F), customers/truck (X), weekly workdays (W)
2. **Calculations:** Number of trucks (N), see Equation (4.1)
3. **Output:** Number of trucks (N)

Program 4.2 Listing:

```
'****************************************************************
********************
'Program 4.2: Calculates number of collection vehicles needed
by a community
'****************************************************************
********************
Public Class Form1

    Private Sub Form1_Load(ByVal sender As System.Object,
            ByVal e As System.EventArgs) Handles MyBase.Load
        Me.Text = "Program 4.2: Calculates number of collection
            vehicles needed by a community"
        Me.FormBorderStyle = BorderStyle.FixedSingle
        Me.MaximizeBox = False

        Label1.Text = "Total number of customers serviced S:"
        Label2.Text = "Collection frequency F (per week):"
        Label3.Text = "Number of customers a single truck can
service
                per day X:"
        Label4.Text = "Number of workdays W (per week):"
        Label5.Text = ""
        Label6.Text = "Formula:" + vbCrLf + "N = (S* F)/(X *
W)"
        Button1.Text = "&Calculate"
        TextBox3.Text = "300"
    End Sub

    Sub calculateResults()
        Dim str As String = ""
        Dim S, F, X, W, N As Double
        S = Val(TextBox1.Text)
        F = Val(TextBox2.Text)
```

94

```vbnet
        X = Val(TextBox3.Text)
        W = Val(TextBox4.Text)

        N = (S * F) / (X * W)
        str = "N = " + N.ToString
        str += vbCrLf + "The community will need " +
            numberInLetters(Math.Ceiling(N)) + " trucks."

        Label5.Text = str
    End Sub


'*********************************************************

'*********************************************************
    'takes a number as integer, and returns it in letters.
    'e.g. 1 -> 'One', 2 -> 'Two' and so one.

'*********************************************************

'*********************************************************
    Function numberInLetters(ByVal n As Integer) As String
        Select Case n
            Case 0 : Return "Zero"
            Case 1 : Return "One"
            Case 2 : Return "Two"
            Case 3 : Return "Three"
            Case 4 : Return "Four"
            Case 5 : Return "Five"
            Case 6 : Return "Six"
            Case 7 : Return "Seven"
            Case 8 : Return "Eight"
            Case 9 : Return "Nine"
            Case 10 : Return "Ten"
            Case 11 : Return "Eleven"
            Case 12 : Return "Twelve"
            Case 13 : Return "Thirteen"
            Case 14 : Return "Fourteen"
            Case 15 : Return "Fifteen"
            Case 16 : Return "Sixteen"
            Case 17 : Return "Seventeen"
            Case 18 : Return "Eighteen"
            Case 19 : Return "Nineteen"
            Case 20 : Return "Twenty"
            Case Else : Return ""
        End Select
```

```
End Function

Private Sub Button1_Click(ByVal sender As System.Object,
        ByVal e As System.EventArgs) Handles
Button1.Click
        calculateResults()
    End Sub
End Class
```

Exercise (4.1)

1) Write briefly about the following agenda concerning solid waste as related to your research project:
   - Process of collection.
   - Sorting of SW components from each other.
   - Responsible personnel for collection and sorting of SW.
   - Preferred method to transport in your city. Give reasons for your answer.
   - Objectives of SW collection.
   - Stages of SW collection.
   - Difference between collection of SW in both rural and urban areas.
   - Appropriate routes for car collecting SW between neighborhoods in the city.
   - Disadvantages and advantages of transfer stations.
   - Transfer station location, divisions available and methods to unload SW trucks.
   - Difference between re-use and recycling.
   - Methods of collecting recyclable materials.
   - Methods of storage of SW in house, apartment, and office. Harmful effects for keeping SW for a long time.
   - Kinds of baskets preferred for storage until transferred.
   - Anti-breeding of trash flies used risks resulting from breeding of blow flies.
   - Appropriate hours of collection of SW and related reasons
   - Obstacles to collect SW in your area. Give most appropriate solutions to improve the situation.

2) Write briefly about oobjectives of municipal solid waste, MSW, collection in Al-Danaha municipality. (B.Sc., UoD, 2012)

3) Write briefly about mmajor factors affecting cost of municipal solid waste, MSW, collection. (B.Sc., UoD, 2012)

4) Write briefly about stages of municipal solid waste, MSW, process collection in an urban area. (B.Sc., UoD, 2012)

5) What innovative methods to finance and fund programs for collection and disposal of municipal solid waste, MSW, would you propose to be adopted in KSA. Outline your reasons for your proposal? (B.Sc., UoD, 2012)

6) Write briefly about role of a transfer station in municipal solid waste, MSW, collection. Indicate when it is preferred to rely on transfer stations. (B.Sc., UoD, 2012)

7) Write briefly about municipal solid waste, MSW, recycling and reuse. (B.Sc., UoD, 2012)

8) Write briefly about objectives of municipal solid waste, MSW, collection. (B.Sc., UoD, 2012)

9) Write briefly about refuse collection phases. **(B.Sc., UoD, 2012)**

10) Indicate how to use the following equation by the mass of materials for estimating the bulk density, ρ_i, of a mixture of solid waste. (B.Sc., UoD, 2013)

$$\rho_i = \frac{\sum\limits_{i=1}^{n}(M_i)}{\sum\limits_{i=1}^{n}\left(\dfrac{M_i}{\rho_i}\right)}$$

Exercise (4.2)

1) A family of four people generates MSW at a rate of 2.5 lb/cap/day and the bulk density of refuse in a typical garbage can is about 250 lb/yd^3. If collection is once a week, how many 30-gallon garbage cans will they need, or the alternative, how many compacted 20-lb blocks would the family produce if they had a home compactor? How many cans would they need in that case? (B.Sc., UoD, 2012)

2) A family of five people generates solid waste at a rate of 2 lb./cap/day. Collection is once a week. How many customers can a 20-yd^3 truck that compacts the refuse to 500 lb/yd^3

collect before it has to make a trip to the landfill? (UoD, B.Sc. 2012)

3) A family of five people generates solid waste at a rate of 2 lb./cap/day and the bulk density of refuse in a typical garbage can is about 200 lb./yd3. If collection is once a week, how many 30-gallon garbage cans will they need? Or the alternative, how many compacted 20-lb blocks would the family produce if they had a home compactor? How many cans would they need in that case? Assume that the bulk density of the compacted refuse is about 1400 lb/yd³ (830 kg/m³). (B.Sc., UoD, 2012)

4) The refuse from a certain area has the following tabulated components and bulk densities: (B.Sc., UoD, 2013)

Component	Percentage (by weight)	Uncompacted bulk density
Cardboard	15	1161 (lb/yd³)
Garden waste	25	4.45 (lb/ft³)
Glass	20	18.45 (lb/ft³)
Miscellaneous paper	40	3.81 (lb/ft³)

Assume that the compaction in the landfill is 1200 lb/yd³.
Estimate percent volume reduction achieved during compaction of the waste using the following equation:

$$\rho_{(A+B+C+D)} = 0.0897 = \frac{15 + 35 + 15 + 35}{\dfrac{15}{0.038} + \dfrac{35}{\rho_{food}} + \dfrac{15}{0.295} + \dfrac{35}{0.061}}$$

Estimate the overall uncompacted bulk density if the miscellaneous paper is removed. Estimate percent volume reduction achieved. Comment on your results (Ans. 0.128, 230 lb/yr³)

5) A family of four people generates solid waste at a rate of 2 lb./cap/day. Collection is once a week. How many customers can a 20-yd³ truck that compact the refuse to 500 lb/yd³ and collect before it has to make a trip to the landfill? (B.Sc., UoD, 2013) (Ans. 179)

6) Assume each household produces 50 lb of refuse per week (as in problem 2). How many customers can a 20-yd³ truck

that compacts the refuse to 600 lb/yd^3 collect before it has to make a trip to the landfill?

7) Assume each household produces 60 lb of refuse per week. How many customers can a 20-yd^3 truck that compacts the refuse to 550 lb/yd^3 collect before it has to make a trip to the landfill? (B.Sc., UoD, 2012)

8) Suppose a crew of two people requires 2 minutes per stop, at which they can service four customers. If each customer generates 50 lb of refuse per week, how many customers can they service if they did not have to go to the landfill?

9) Suppose a crew of two people requires 3 minutes per stop, at which they can service four customers. If each customer generates 50 lb of refuse per week, how many customers can they service if they did not have to go to the landfill?

10) Suppose a crew of two people requires 2 minutes per stop, at which they can service five customers. Assume that a working day is 8 hours. How many customers can they service if they did not have to go to the landfill? (B.Sc., UoD, 2012)

11) A community of 1200 homes cannot pay for the initial and operating costs of the recycling collection vehicles that were to be used. Instead, residents are to haul recycling containers to a drop-off center operated by the community. Calculate the number of vehicles from which recyclable materials must be unloaded per hour at the recycling drop-off center. Assume the center is open for eight hours per day, two days per week, and that 40 percent of the residents will deliver recycling containers. Also assume that 75 percent of the participants will take their separated materials to the drop-off center once per week and that the remaining 25 percent of the participants will bring their separated materials to the drop-off center once every two weeks.

12) A truck is found to be able to service customers at a rate of 2 customers per minute. If they find that the actual time they spend on collection is 5 hours, how many customers can be served per day?

99

13) A truck is found to be able to service customers at a rate of 1.5 customers per minute. If they find that the actual time they spend on collection is 5 hours, how many customers can be served per day?

14) Calculate the number of collection vehicles a community would need if it has a total of 4000 services (customers) that are to be collected once per week.

15) Calculate the number of collection vehicles a community would need if it has a total of 6000 services (customers) that are to be collected once per week.

16) The number of collection vehicles a community would need for solid waste collection in a certain municipality is 6 trucks. Find total number of services (customers) that are to be collected once per week during working days in a Al-Doha in KSA. (Realistically, most trucks can service only about 200 to 300 customers before the truck is full and a trip to the landfill is necessary). (B.Sc., UoD, 2012)

17) Define the terms contained in the equation used to determine maximum volume that can be squeezed by a crushing cylinder (roll) $C = k.v.D.L.s.r$

Find capacity in (tons / hour) of a crushing cylinder noting that speed of cylinder is 60 cycles per minute, cylinder diameter of 20 cm, its length 0.35 meters, and taking density of material of 2.6 g/cm3 and distance between discs of 5 mm. Assume crushing roll constant = 60 when taking units and dimensions described in this equation. (UoD, B.Sc. 2012).

Chapter Five

Processing Solid Waste and Material Separation

5.1 Background

MSW refuse is a heterogeneous material with unpredictable and time-variable characteristics and amounts. These conditions interfere negatively with design MSW processing and materials recovery facilities, MRFs, when attempting to utilize the material in producing a desirable product. Some types of MSW refuse are easily processed, yet certain other types are difficult and/or dangerous to handle. MSW processing operations incorporate many interlinked factors at operational, organizational and safety levels. Therefore, the material ought to be designed for extraordinary contingencies. Such a requirement often results in overdesign and underutilization of resources when processing all of the feed material.

5.2 Primary MSW treatment operations

Primary treatment operations to prepare MSW and garbage in a sustainable management system format target increasing efficiency of operation, extracting useful materials and resources and restoring products and energy. This could be achieved through: size reduction by mechanical means (e.g. compaction or fragmentation) or chemical means (e.g. incineration and burning); or by automatic and mechanical separation of components; or extraction of moisture content and drying.

Basic types of conveyors used primarily to move MSW and refuse include: rubber-belted conveyors, live bottom feeders and pneumatic conveyors. Other conveyors used to feed or meter refuse to a load-sensitive device (such as a combustor) incorporate: vibratory feeders, screw feeders and drag chains.

Power requirements of belt conveyors can be estimated by a number of empirical equations. Equation 5.1 represent the power necessary to move a load horizontally and vertically together with power loss due to friction.

$$\text{Horsepower} = \frac{LSF}{1000} + \frac{LTC}{990} + \frac{TH}{990} + P \qquad (5.1)$$

Where
L = Length of conveyor belt, ft
S = Speed of belt, ft/min
F = Speed factor, dimensionless
T = Capacity, tons/h
C = Idle resistance factor, dimensionless
H = Lift, ft
P = Pulley friction, horsepower

Screw conveyors are used to meter shredded refuse into a furnace. The volume of material moved by a screw conveyor in a flooded condition can be estimated by equation 5.2.
Q = CNRV   (5.2)

Where
Q = Delivery of refuse, m^3/min.
C = Efficiency factor.
N = Number of conveyor leads or number of blades that are wrapped around the conveyor hub.
R = Rotational speed of screw, rpm.
V = Volume of refuse between each pitch, m^3

Crushing devices or rolls break fragile materials such as glass, while unfolding rout iron such as iron cans, which would facilitate separation by sieving. Crushing rolls strongly hold the raw material entering between two rollers operating in opposite directions. The maximum volume that can be squeezed by crushing cylinders can be estimated by equation 5.3.

C = k.v.D.L.s.ρ   (5.3)

Where:
C = Capacity, tons/hour.
k = Constant, dimensionless = 60 When taking units and dimensions described in this equation.
v = Speed of cylinders, rpm.
D = Cylinder diameter, m.
L = Length of cylinder, m.
s = Distance of separation between cylinders (discs), m.
ρ = Density of the material, g/cm^3.

Example (5.1)

a) Find capacity of a crushing cylinder noting that speed of cylinder is 4200 cycles per hour, cylinder diameter of 250 mm, its length 45 cm, and taking density of material of 2500 kg/m^3 and distance between discs of 5 mm.
b) Write a computer program to calculate the capacity of a crushing cylinder given speed of cylinder, cylinder diameter, its length, density of material and distance between discs.
c) Verify your program by solving example 5.1.

Solution

1) Given: v = 4200 cycles/hr = 4200/60 cycles/min, D = 250 mm = 20 cm, L = 0.45 meters, ρ = 2500/1000 = 2.5 g/cm^3, s = 5 mm.
2) Then, capacity, C = k.v.D.L.s.r
   C = k.v.D.L.s.r = 60 x (4200/60) rpm x 0.25 m x 0.45 m x 0.005 m x 2.5 (g/cm^3) = 7.1 tones/hr.

Program 5.1 Algorithm: capacity of a crushing cylinder

1. **Inputs:** Cylinder speed (v, cyc/hr), cylinder diameter (D, cm), cylinder length (L, m), material density (ρ), disc distance (s, mm)
2. **Calculations:** See Equation (5.3)
3. **Output:** Capacity of the crushing cylinder (C)

Program 5.1 Listing:

```
'****************************************************************
'Program 5.1: Calculates capacity of a crushing cylinder
'****************************************************************
Public Class Form1

    Private Sub Form1_Load(ByVal sender As System.Object,
            ByVal e As System.EventArgs) Handles MyBase.Load
        Me.Text = "Program 4.1: Calculates capacity of a
                        crushing cylinder "
        Me.MaximizeBox = False
        Me.FormBorderStyle =
            Windows.Forms.FormBorderStyle.FixedSingle

        Label1.Text = "Formula:" + vbCrLf
        Label1.Text += "C = k.v.D.L.s.rho"

        Label2.Text = "Speed of cylinders (v), cycles/hr:"
        Label3.Text = "Cylinder diameter (D), mm:"
        Label4.Text = "Length of cylinder (L), cm:"
        Label5.Text = "Distance of separation between discs
(s), mm:"
        Label6.Text = "Density of the material (rho), kg/m3:"
        Label7.Text = ""
        Button1.Text = "&Calculate"
    End Sub

    Sub calculateResults()
        Dim C, k, v, D, L, s, rho As Double
        v = Val(TextBox1.Text)
        D = Val(TextBox2.Text)
        L = Val(TextBox3.Text)
        s = Val(TextBox4.Text)
        rho = Val(TextBox5.Text)

        k = 60
        v = v / 60   'convert to cycles/min
        D = D / 1000    'convert to m.
        L = L / 100 'convert to m.
        s = s / 1000    'convert to m.
        rho = rho / 1000    'convert to g/cm3

        C = k * v * D * L * s * rho
        Label7.Text = "Capacity, C = " + C.ToString + "
tones/hr"
    End Sub
```

```
Private Sub Button1_Click(ByVal sender As System.Object,
        ByVal e As System.EventArgs) Handles
Button1.Click
        calculateResults()
    End Sub
End Class
```

5.3 MSW separation

In separating various pure materials from a mixture, the separation can be either binary (two output streams) or polynary (more than two output streams). A binary separator is designed to extract one type of material from a waste stream (e.g. magnet drawing off ferrous materials as the desired output or product or extract). A polynary separator separates components from each other through multiple paths (e.g. screen with a series of different sized holes, producing several products).

Suppose in a binary separator the input stream is composed of a mixture of x and y to be separated (see figure 5.1). The mass per time (e.g., tons/hour) of x and y fed to the separator is x_0 and y_0, respectively. The mass per time of x and y exiting in the first output stream is x_1 and y_1, and the second output stream is x_2 and y_2. The device separates the x into the first output stream and y into the second. The effectiveness of the separation then can be expressed in terms of recovery.

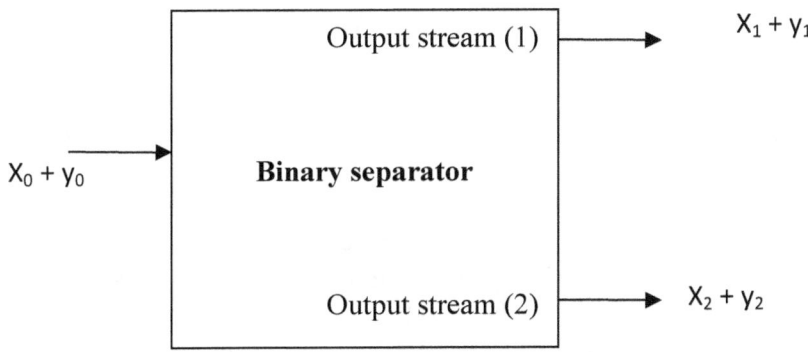

Figure 5.1: Binary separator.

105

The recovery of component x in the first output stream is Rx1, defined as shown in equation 5.4.

$$R_{x_1} = \frac{x_1}{x_o} \times 100 \tag{5.4}$$

Where:
R_{x1} = Recovery of component x in the first output stream (1).
x_1 = First component emerging of the first output stream (1), mass/time.
x_o = x component entering to the binary separator, mass/time.

Similarly, the recovery of y in the second output stream may be found from equation 5.5.

$$R_{y2} = \frac{y_2}{y_o} \times 100 \tag{5.5}$$

Purity of output can be determined from equation 5.6.

$$P_{x_1} = \frac{x_1}{(x_1 + x_2)} \times 100 \tag{5.6}$$

Where:
P_{x1} = Purity of the first output stream in terms of x, which is expressed as a percentage.

Similarly, the purity of the second output stream in terms of y is as presented in equation 5.7.

$$P_{y_2} = \frac{y_2}{(x_2 + y_2)} \times 100 \tag{5.7}$$

Overall recovery is useful only for process design, such as sizing conveyor belts. Nevertheless, it is not a measure of separation effectiveness. It should not be used in describing the operation of materials separation. Overall recovery may be defined as in equation 5.8.

$$OR_{x,y} = \left(\frac{x_1 + y_1}{x_0 + y_0}\right) \times 100 \tag{5.8}$$

Where:

$OR_{x,y}$ = Overall recovery for components x and y.

Rietema separation effectiveness measure for a binary separation with input of x_0 and y_0 yields equation 5.9.

$$E_{x,y} = 100 \left[\frac{x_1}{x_0} - \frac{y_1}{y_0}\right] = 100 \left[\frac{x_2}{x_0} - \frac{y_2}{y_0}\right] \tag{5.9}$$

Worrell and Stessel equation for finding the separation performance of a binary separator is reflected in equation 5.10.

$$E_{x,y} = \left[\frac{x_1}{x_0} \frac{y_2}{y_0}\right]^{1/2} \times 100 \tag{5.10}$$

Example (5.2)

a) A binary separator has a feed rate of 1 ton/h. It is operated so that during any 1 hour, 700 kilograms reports as output (1) and 300 kg as output (2). Of the 700 kg the x constituent is 650 kg, while 70 kg of x ends up in output 2. Find the recoveries and the effectiveness of separation using different methods.

$$T = 1 \text{ hr}$$

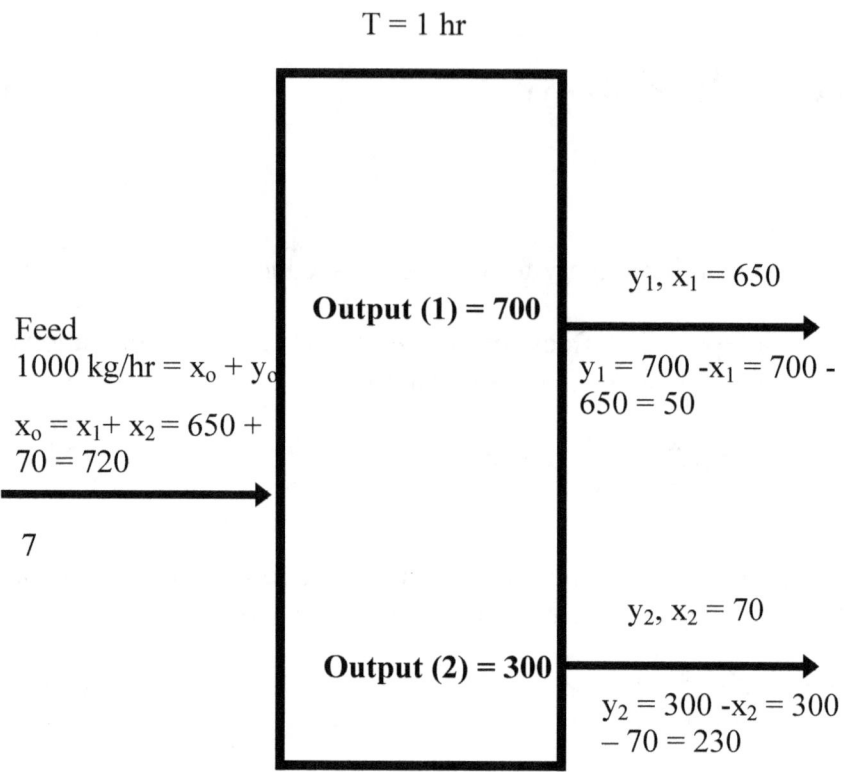

Output (1) = 700

$y_1, x_1 = 650$

Feed
$1000 \text{ kg/hr} = x_o + y_o$

$y_1 = 700 - x_1 = 700 - 650 = 50$

$x_o = x_1 + x_2 = 650 + 70 = 720$

7

$y_2, x_2 = 70$

Output (2) = 300

$y_2 = 300 - x_2 = 300 - 70 = 230$

b) Write a computer program to calculate the recoveries and effectiveness of separation a binary separator using different methods given: hourly feed rate, operation time, mass produced as output (1) and output (2) and their respective x constituent.

c) Verify your program by solving example 5.2.

Solution

1) Given: Data: $x_1 = 650$ kg, $x_o + y_o = 1000$ kg (1 ton), $x_2 = 70$ kg. (See figure).

2) From data find $x_o = x_1 + x_2 = 650 + 70 = 720$ kg,

3) Then, find y_o = total value - x_o = 1000 - 720 = 280 kg.

4) From track (1) the value of $y_1 = 700 - x_1 = 700 - 650 = 50$ kg.

5) From track (2) the value of $y_2 = 300 - x_2 = 300 - 70 = 230$ kg.

6) Find the recovery of x in the first output from equation

$$R_{x_1} = \frac{x_1}{x_0} \times 100 = \frac{650}{720} \times 100 = 90$$

7) Find purity of this output stream from equation :

8)
$$P_{x_1} = \frac{x_1}{x_1 + y_1} \times 100 = \frac{650}{650 + 50} \times 100 = 93$$

Using Rietema's definition of effectiveness

9)
$$E_{x,y} = 100 \left[\frac{x_1}{x_o} - \frac{y_1}{y_o} \right] = 100 \left[\frac{650}{720} - \frac{50}{280} \right] = 72$$

and according to Worrell-Stessel effectiveness equation,

$$E_{x,y} = \sqrt{\frac{x_1}{x_o} \frac{y_2}{y_o}} \times 100 = \sqrt{\frac{650}{720} \frac{230}{280}} \times 100 = 86$$

A binary separator has a feed rate of 1 ton/h. It is operated so that during any 1 hour, 700 kilograms reports as output (1) and 300 kg as output (2). Of the 700 kg the x constituent is 650 kg, while 70 kg of x ends up in output 2. Find the recoveries and the effectiveness of separation using different methods.

Program 5.2 Algorithm: recoveries and effectiveness of separation a binary separator using different methods

1. **Inputs:** x_1, x_2, $(x_o + y_o)$, output1, output2
2. **Calculations:** y_1, y_2, See Equations (5.5), (5.5), and (5.10)
3. **Output:** (1) The recoveries, and (2) the effectiveness of separation (using different methods)

Program 5.2 Listing:

```vb
'***************************************************************
*******
'Program 5.2: Calculates the recoveries and the effectiveness
'of separation of a binary separator.
'***************************************************************
*******
Imports System.Math

Public Class Form1

    Private Sub Form1_Load(ByVal sender As System.Object,
            ByVal e As System.EventArgs) Handles MyBase.Load
        Me.Text = "Program 4.2:"
        Me.FormBorderStyle =
            Windows.Forms.FormBorderStyle.FixedSingle
        Me.MaximizeBox = False

        Label1.Text = "Program 5.2: Calculates the recoveries
and the
                        effectiveness "
        Label1.Text += vbCrLf + "of separation of a binary
separator."

        Label2.Text = "Feed rate (kg/hour):"
        Label3.Text = "Hourly output (1) (kg):"
        Label4.Text = "Hourly output (2) (kg):"
        Label5.Text = "The x constituent (x1) (kg):"
        Label6.Text = "The amount of x ending up in output2
(x2) (kg):"
        Label7.Text = ""
        button1.text = "&Calculate"
    End Sub

    Sub calculateResults()
        Dim feed, x0, x1, x2, y0, y1, y2, Rx, Px, Exy, Exy2 As
Double
        Dim out1, out2 As Double

        feed = Val(TextBox1.Text)
        out1 = Val(TextBox2.Text)
        out2 = Val(TextBox3.Text)
        x1 = Val(TextBox4.Text)
        x2 = Val(TextBox5.Text)

        x0 = x1 + x2
```

111

```vbnet
        y0 = feed - x0
        y1 = out1 - x1
        y2 = out2 - x2
        'Find the recovery of x in the first output:
        Rx = (x1 / x0) * 100
        'Find purity of this output stream:
        Px = x1 / (x1 + y1) * 100
        'Using Rietema's definition of effectiveness:
        Exy = 100 * abs((x1 / x0) - (y1 / y0))
        'Worrell-Stessel effectiveness equation:
        Exy2 = sqrt(abs((x1 / x0) * (y2 / y0))) * 100

        Dim str As String
        str = "x0 = x1 + x2 = " + fN(x0) + "kg"
        str += vbCrLf + "y0 = total - x0 = " + fN(y0) + "kg"
        str += vbCrLf + "y1 = output1 - x1 = " + fN(y1) + "kg"
        str += vbCrLf + "y2 = output2 - x2 = " + fN(y2) + "kg"
        str += vbCrLf + "Recovery of x in the first output Rx1
= "
                + fN(Rx)
        str += vbCrLf + "Purity of output stream Px1 = " +
fN(Px)
        str += vbCrLf + "Effectiveness (using Rietema's
definition) = "
                + fN(Exy)
        str += vbCrLf + "Effectiveness (using Worrell-Stessel
eqn.) = "
                + fN(Exy2)

        Label7.Text = str
    End Sub

    Function fN(ByVal n As Double) As String
        'formats a number 'n' as ##.# and returns result as
string
        Return Format(n, "##.#")
    End Function
    Private Sub Button1_Click(ByVal sender As System.Object,
            ByVal e As System.EventArgs) Handles
Button1.Click
        calculateResults()
    End Sub
End Class
```

Exercise (5.1)

1) Write briefly about primary treatment systems to prepare municipal solid waste, MSW. (B.Sc., UoD, 2012)
2) Differentiate between recovery, purity, Rietema effectiveness and Worrell & Stessel equation as means of effectiveness separation expressions. (B.Sc., UoD, 2013)

Exercise (5.2)

1) Find capacity of a crushing cylinder noting that speed of cylinder is 50 cycles per minute, cylinder diameter of 20 cm, its length 0.6 meters, and taking density of material of 2.3 g/cm³ and distance between discs of 5 mm.
2) Find capacity of a crushing cylinder noting that speed of cylinder is 60 cycles per minute, cylinder diameter of 30 cm, its length 0.5 meters, and taking density of material of 2.5 g/cm³ and distance between discs of 5 mm.
3) Determine the capacity (in tons/hr) of a crushing cylinder noting that speed of cylinder is 70 cycles per minute, cylinder diameter of 25 cm, its length 0.3 meters, and taking density of material of 2.5 g/cm³ and distance between discs of 5 mm. Assume crushing roll constant = 60 when taking units and dimensions described in this equation. (B.Sc., UoD, 2013) (Ans. 3.9375 tones/hr)
4) A binary separator has a feed rate of 1 tonne/h. It is operated so that during any 1 hour, 800 kg reports as output 1 and 200 kg as output 2. Of the 800 kg, the x constituent is 750 kg, while 90 kg of x ends up in output 2. Calculate the recoveries and the effectiveness of the separation using different methods.
5) A binary separator has a feed rate of 1 tonne/h. It is operated so that during any 1 hour, 700 kg reports as output 1 and 300 kg as output 2. Of the 700 kg, the x constituent is 650 kg, while 60 kg of x ends up in output 2. Calculate the recoveries and the effectiveness of the separation using the methods of recovery, purity, Rietema effectiveness and Worrell & Stessel equation. Comment on your results. (B.Sc., UoD, 2013) (Ans. 92, 93, 74, 87%)
6) Assuming that the drag coefficient is 2.4, calculate the air velocity necessary to suspend 1.5 cm (screened) particles of

shredded aluminum. Note that $\rho_s = 1.1$ g/cm^3 & $\rho = 0.001$ g/cm^3. (Hint: Use Newton's law to compute velocity) (B.Sc., UoD, 2013) (Ans. 947 cm/s)

Chapter Six

Sanitary Landfill

6.1 Overall view

A sanitary landfill is an engineered method for land disposal of solid or hazardous wastes in a manner that protects the environment. Within the landfill biological, chemical, and physical processes occur biodegrading wastes and resulting in the production of leachate and gaseous substances. Leachate production and quantity can be estimated using empirical data or a water balance technique.

6.2 Water balance system in landfill

Water balance system in the landfill facilitates estimating amount of percolating water production by establishing a mass balance among precipitation, evapotranspiration, surface runoff, and soil moisture storage as presented in equation 6.1.

$$C = P (1 - r) - S - E \qquad\qquad (6.1)$$

Where:
C = Total amount percolating within the top layer of soil, mm/year.
P = Precipitation, mm/year.
r = Coefficient of runoff (can be estimated for different types of soil).
S = Storage in the soil or solid waste, mm/year.
E = Evapotranspiration, mm/year.

5.3 Landfill gas production

Gas production over time may be estimated from the EPA LandGEM model[10] based on equation 6.2.

$$\cdot\, Q_T = \sum_{i=1}^{n} 2\, k L_o M_i e^{-kt_i} \ \rule{2cm}{0.4pt}$$

[10]   This model can be downloaded at http://www.epa.gov/ttn/catc/products.html#software.

$$(6.2)$$

Where:

Q_T = Total gas emission rate from a landfill, volume/time.

n = Total time periods of waste placement.

k = Landfill gas emission constant, $time^{-1}$.

L_0 = Methane generation potential, volume/mass of waste.

t_i = Age of the ith section of waste, time.

M_i = Mass of wet waste, placed at time i.

Example (6.1)

a) A landfill cell is open for four years, receiving 90,000 tons of waste per year (recall that 1 ton = 1000 kilograms). Find the peak gas production for the first year if the landfill gas emission constant is 0.03 per year, and the methane generation potential is 140 m^3/ton.

b) Write a computer program to calculate the peak gas production for the first year from a certain landfill cell given cell opening period, annual tons of waste received by the site, the landfill gas emission constant, and the methane generation potential.

c) Verify your program by solving example 6.1.

Solution

1) Given: Mi =90000 t/yr, k = 0.03, Lo = 140, ti = 1

$$Q_T = \sum_{i=1}^{n} 2k \, L_o M_i \, e^{-kti}$$

2) For the first year,

$Q_T = 2 \, (0.03) \, (140) \, (90000)(e^{-0.0307(1)}) = 733657 \ m^3$

Program 6.1 Algorithm: peak gas production from a certain landfill

1. **Inputs:** Mi (ton/yr), Lo (m3/ton), k (constant)
2. **Calculations:** QT, See Equation (6.2)
3. **Output:** The peak gas production for the first year (QT, m^3)

Program 6.1 Listing:

```
'************************************************************
***********
'Program 6.1: Calculates the peak gas production for a number
of years
'************************************************************
***********
Public Class Form1
    Dim Mi(), k(), Lo() As Double
    Dim Qt, t As Double
    Dim n As Integer

    Private Sub Form1_Load(ByVal sender As System.Object,
            ByVal e As System.EventArgs) Handles MyBase.Load
        Me.Text = "Program 6.1: Calculates the peak gas
production for
                   a number of years"
        Me.FormBorderStyle =
                   Windows.Forms.FormBorderStyle.FixedSingle
        Me.MaximizeBox = False

        Label1.Text = "For each year, enter the annual waste
received
                   (Mi), gas emission constant (k),"
        Label1.Text += vbCrLf + "and the methane generation
potential
                   (Lo):"

        DataGridView1.Columns.Clear()
        DataGridView1.Columns.Add("MCol", "Mi (tones/yr)")
        DataGridView1.Columns.Add("KCol", "K (time-1)")
        DataGridView1.Columns.Add("LCol", "Lo (m3/ton)")

        Label2.Text = ""
        Button1.Text = "&Calculate"
    End Sub

    Private Sub Button1_Click(ByVal sender As System.Object,
            ByVal e As System.EventArgs) Handles
Button1.Click
        calculateResults()
    End Sub

    Sub calculateResults()
```

```
        n = DataGridView1.Rows.Count - 1
        If n <= 0 Then
            MsgBox("Enter at least one period's data!",
vbOKOnly,
                    "Error")
            Exit Sub
        End If


        '*********************************************
        'Formula is:
        '          n
        '     QT = SUM 2k * Lo * Mi * e^(-kti)
        '          i=1
        '*********************************************
        Qt = 0
        t = 0
        ReDim Mi(n), k(n), Lo(n)
        For i = 0 To n - 1
            Mi(i) =
Val(DataGridView1.Rows(i).Cells("MCol").Value)
            k(i) =
Val(DataGridView1.Rows(i).Cells("KCol").Value)
            Lo(i) =
Val(DataGridView1.Rows(i).Cells("LCol").Value)
            t += 1
            Qt += 2 * k(i) * Lo(i) * Mi(i) * (Math.E ^ (-k(i) *
t))
        Next


        If n = 1 Then
            Label2.Text = "For one year, QT = " + Qt.ToString +
" m3"
        Else
            Label2.Text = "For " + n.ToString + " years, QT = "
+ Qt.ToString + " m3"
        End If
    End Sub
End Class
```

Example (6.2)

Use the EPA Landfill Gas Emissions model (LandGEM) model used for estimating methane production from a sanitary landfill. A landfill cell is open for three years; receiving 200,000 tonnes of waste per year (recall that 1 tonne 1000 kg). Calculate the peak gas production if the landfill-gas emission constant is 0.05 yr^{-1} & the methane generation potential is 170 m^3/tonne. (Hint: for the second year this waste produces less gas, but the next new layer produces more, and the two layers are added to yield the total gas production for the second year.) (B.Sc., UoD, 2013)

Solution

Q_T = Total gas emission rate from a landfill, volume/time

n = Total time periods of waste placement

k = Landfill gas emission constant, time1

L_o = Methane generation potential, volume/mass of waste

t_i = Age of the ith section of waste, time

M_i = Mass of wet waste, placed at time i

Given: M_i =200,000 t/yr, k = 0.05 yr^{-1}, L_o = 170 m^3/tonne, t_i = 1 yr

$$Q_T = \sum_{i=1}^{n} 2k\, L_o M_i\, e^{-kti}$$

For the first year,

$QT_1 = 2\,(0.05)\,(170)\,(200,000)(e^{-0.05*1}) = 3,234,180\ m^3$

Determine gas produced for other years as shown in table below

T	mass, M	mass, M	constant, k	emission, L	total gass, Q
1	200000	200000	0.05	170	3234180
2	200000	400000	0.05	170	6152894
3	200000	600000	0.05	170	8779221
4	200000	497100	0.05	170	6918848
5	200000	497100	0.05	170	6581412
6	200000	497100	0.05	170	6260433
7	200000	497100	0.05	170	5955108
8	200000	497100	0.05	170	5664674
9	200000	497100	0.05	170	5388404
10	200000	497100	0.05	170	5125609

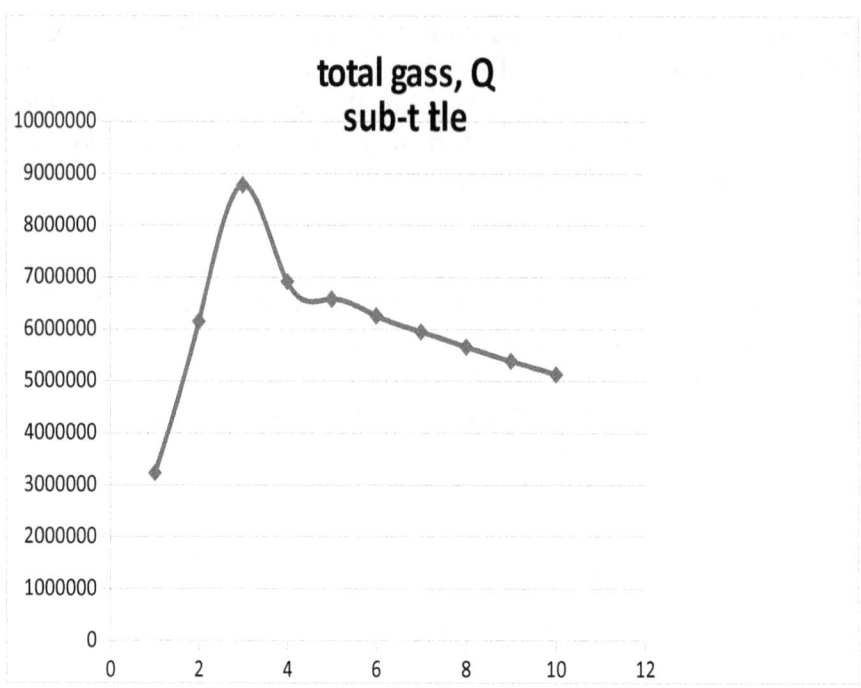

6.4 Depth of leachate in the lining

Depth of leachate in the lining can be estimated using Darcy's law and continuity equation, depending on: rates of filtration, permeability of drainage materials, distance from discharge tube, slope of drainage system. Equation 6.3 illustrates depth of leachate estimates.

$$Y_{max} = \frac{P}{2}\left(\frac{q}{k}\right)\left[\frac{k.\tan^2\alpha}{q} + 1 - \frac{k.\tan\alpha}{q}\left(\tan^2\alpha + \frac{q}{k}\right)^{\frac{1}{2}}\right] \qquad (6.3)$$

Where:
Y_{max} = Maximum saturated depth over the liner, ft
P = Distance between collection pipes, ft
q = Vertical inflow (infiltration), defined in this equation as from a 25-year, 24-hour storm, ft/day
α = Inclination of liner from horizontal, degrees

120

K = Hydraulic conductivity of the drainage layer, ft/day

Example (6.3)

The maximum design depth of leachate above lining in a sanitary landfill is 16 cm and the distance between leachate collection tubes is 19 meters. The hydraulic conductivity is 0.02 cm/s for a slope of discharge 2 percent. Estimate the design storm (vertical flow) in cm/s using a coarse discharge material & assuming that rain water from 25 years & a storm entering the leachate drainage system of a 24-hour. (B.Sc., UoD, 2013)

Solution

Given: Y_{max} = 16 cm, P = 1900 cm, k = 0.02 cm/s, q = 0.0003 cm/s, k = ? cm/s, tan a = 2/100 = 0.02

$$Y_{max} = \frac{P}{2}\left(\frac{q}{k}\right)\left[\frac{k.\tan^2 a}{q} + 1 - \frac{k.\tan a}{q}\left(\tan^2 a + \frac{q}{k}\right)^{\frac{1}{2}}\right]$$

$Y_{max} =$

$$\frac{1900}{2}\left(\frac{q}{0.02}\right)\left[\frac{0.02*(0.02^2)}{q} + 1 - \frac{0.02*(0.02)}{q}\left(0.02^2 + \frac{q}{0.02}\right)^{\frac{1}{2}}\right]$$

$$= 16$$

By trial and error q = 0.00038 cm/s. (otherwise use appropriate software to solve the problem such as: http://www.wolframalpha.com/)

5.5 Sanitary landfill area

Needed land area can be found from estimates of required volume from equation 6.4.

$$V = \frac{W}{\rho}\left(1 - \frac{x}{100}\right) + v_r \qquad (6.4)$$

Where:
V = Volume of sanitary landfill area.
W = Weight of SW to be buried.

121

ρ = Average density of SW and garbage.

x = Percentage of compressed SW volume,%

Vr = Volume of a layer of coverage required (thickness of 15 to 30 cm for medium layers, temporary edge and front and overhead slope, and at least 60 cm in the final layer) and that this volume ranges between 17% of volume of SW for deep burial to 33% for surface burial and in average 25 per cent.

Average sanitary landfill volume can be estimated as shown in equation 6.5.

$$V = 1.25 \frac{W}{\rho}\left(1 - \frac{x}{100}\right) \qquad (6.5)$$

Exercise (6.1)

1) Define a sanitary landfill. (B.Sc., UoD, 2013)
2) Define the terms used in the following equation for estimating depth of leachate in the lining of a landfill. (B.Sc., UoD, 2013)

$$Y_{max} = \frac{P}{2}\left(\frac{q}{k}\right)\left[\frac{k.\tan^2\alpha}{q} + 1 - \frac{k.\tan\alpha}{q}\left(\tan^2\alpha + \frac{q}{k}\right)^{\frac{1}{2}}\right]$$

3) What are the main factors affecting depth of leachate in the lining of a landfill. (B.Sc., UoD, 2013)

Exercise (6.2)

1) Find the percolation of water through a sanitary landfill assuming the amount of rainfall is 1200 mm per year, and transpiration 480 mm/year. Assuming a runoff coefficient of 0.14.
2) Estimate the percolation of water through a landfill 15 m deep, with a 1.2 m cover of sandy loam soil. Use the following data:
   - P = 1000 mm/yr
   - R = 0.12
   - E = 600 mm/yr
   - Soil field capacity, F_s = 200 mm/m
   - Refuse field capacity, F_r = 300 mm/m, as packed.

3) Find the percolation of water through a sanitary landfill assuming the amount of rainfall is 1500 mm per year, and transpiration 600 mm/year. Assuming a runoff coefficient of 0.13.

4) A landfill cell is open for three years, receiving 180,000 tonnes[11] of waste per year. Calculate the peak gas production in the first year if the landfill-gas emission constant is 0.03 yr^{-1} and the methane generation potential is 150 m^3/tonne.

5) Use the EPA Landfill Gas Emissions model, LandGEM model used for estimating methane production from a sanitary landfill.

$$Q_T = \sum_{i=1}^{n} 2k \, L_o M_i e^{-kti}$$

A landfill cell is open for three years; receiving 150,000 tonnes of waste per year (recall that 1 tonne 1000 kg). Calculate the peak gas production if the landfill-gas emission constant is 0.031 yr^{-1} & the methane generation potential is 120 m^3/tonne.

(Hint: for the second year this waste produces less gas, but the next new layer produces more, and the two layers are added to yield the total gas production for the second year.) (B.Sc., UoD, 2013) (Ans. 1,081,935, 3050676 m^3)

6) Find maximum design depth above lining noting that distance between leachate collection tubes is 12 meters. Using a coarse discharge material and assuming that rain water from 25 years and a storm entering the leachate drainage system of a 24-hour, design storm (vertical flow) = 0.018 cm/min, and hydraulic conductivity 0.015 cm/s, and slope of discharge 1.5 percent.

7) Determine the spacing between pipes in a leachate collection system using granular drainage material and the following properties. Assume that in the most conservative design all stormwater from a 25-year, 24-hour storm enters the leachate collection system.
   - Design storm (25 years, 24 hours) = 8.2 in = 0.015 cm/min
   - Hydraulic conductivity = 0.01 cm/s

---

[11]   1 tonne = 1000 kg

- Drainage slope = 2%
- Maximum design depth on liner = 16 cm.

8) The maximum design depth above lining in a sanitary landfill is 15.8 cm and the distance between leachate collection tubes is 18 meters. Using a coarse discharge material & assuming that rain water from 25 years & a storm entering the leachate drainage system of a 24-hour, design storm (vertical flow) = 0.0003 cm/s, compute the hydraulic conductivity in cm/s for a slope of discharge 2 percent. (B.Sc., UoD, 2013) (Ans. 0.015 cm/s)

Chapter Seven

Biochemical Processes, Combustion and Energy Recovery

7.1 Background

Municipal solid waste, MSW, components reliable for bioconversion processes are: garbage (food waste), paper products, and yard wastes for their cellulose content. All methods of biochemical conversion (anaerobic digestion, composting) use the organic fraction of refuse. Decay of organic matter under anaerobic conditions produces end-products that include gases such as: methane (CH_4), carbon dioxide (CO_2), small amounts of hydrogen sulfide (H_2S), ammonia (NH_3), and a few others

Ideally, production of methane and carbon dioxide can be calculated using equation 7.1 if chemical composition of a material is known.

$$C_aH_bO_cN_d + \left(\frac{4a-b-2c+3d}{4}\right)H_2O$$

$$\rightarrow \left(\frac{4a+b-2c-3d}{8}\right)CH_4 + \left(\frac{4a-b+2c+3d}{8}\right)CO_2 + dN\,H_3 \qquad (7.1)$$

Example (7.1)

a) Estimate the production of CO_2 and CH_4 during the anaerobic decomposition of Acetic acid, CH_3COOH.

b) Write a computer program to estimate the production of CO_2 and CH_4 during the anaerobic decomposition of a certain organic compound.

c) Verify your program by solving example 7.1.

Solution

1) Given: Reactants and products.

125

2) The formula for acetic acid is CH_3COOH

$$C_aH_bO_cN_d+ \left(\frac{4a-b-2c+3d}{4}\right)H_2O$$

¿

$$\rightarrow \left(\frac{4a+b-2c-3d}{8}\right)CH_4+ \left(\frac{4a-b+2c+3d}{8}\right)CO_2+dN\,H_3¿$$

3) The formula for acetic acid is CH_3COOH, hence by above equation

A = 2, b = 4, c = 2 and d = 0

$CH_3COOH + \{(8-4-4+0)/4\}H_2O = \{(8+4-4-0)/8\} CH_4 + \{8-4+4+0)/8\} CO_2$

$CH_3COOH = CH_4 + CO_2$

4) Note that the equation balances.
5) The molecular weight are 58 = 1(16) + 1(44). Hence, I kg of acetic acid produces 0.27 kg CH_4 and 0.73 kg CO_2.
6) **Recalling** that 1 gm. Molecular weight of a gas at standard temperature and pressure occupies 22.4 liters, the production of CO_2 and CH_4 from one kg of acetic acid is liters each of methane and carbon dioxide.

Program 7.1 Algorithm: production of CO_2 and CH_4 during anaerobic decomposition of an organic compound in a landfill

1. **Inputs:** Equation (reactants, products)
2. **Calculations:** See Equation (7.1)
3. **Output:** The estimated production of CO_2 and CH_4

Program 7.1 Listing:

```
'****************************************************************
'Program 7.1: Estimates the production of CO2 and CH4
'****************************************************************
Imports System.Math

Public Class Form1
    Dim a, b, c, d As Double
    Dim H2O, CH4, CO2, NH3
```

126

```
    Private Sub Form1_Load(ByVal sender As System.Object,
            ByVal e As System.EventArgs) Handles MyBase.Load

        Me.Text = "Program 6.1: Estimates the production of CO2
and
            CH4"
        Me.FormBorderStyle =
            Windows.Forms.FormBorderStyle.FixedSingle
        Me.MaximizeBox = False

        Label1.Text = "Enter the chemical formula of the
material, e.g.
                    Acetic acid is CH3COOH"
        Label2.Text = ""
        Button1.Text = "&Calculate"
    End Sub

    Sub calculateResults()
        readChemicalFormula()

        H2O = ((4 * a) - b - (2 * c) + (3 * d)) / 4
        CH4 = ((4 * a) + b - (2 * c) - (3 * d)) / 8
        CO2 = ((4 * a) - b + (2 * c) + (3 * d)) / 8
        NH3 = d

        H2O = abs(H2O)
        CH4 = abs(CH4)
        CO2 = abs(CO2)
        NH3 = abs(NH3)
        Label2.Text = "a=" + a.ToString + ", b=" + b.ToString +
", c="
                    + c.ToString + ", d=" + d.ToString
        Label2.Text += vbCrLf + TextBox1.Text + " + " +
H2O.ToString
                    + "H2O = " + CH4.ToString + "CH4 + " _
                    + CO2.ToString + "CO2 + "
                    + NH3.ToString + "NH3"

        Dim molCH4, molCO2, totMol
        molCH4 = CH4 * 16    'molecular wt of CH4 is 16
        molCO2 = CO2 * 44    'molecular wt of CO2 is 44
        totMol = molCH4 + molCO2
        molCH4 = molCH4 / totMol
        molCO2 = molCO2 / totMol
        Label2.Text += vbCrLf + vbCrLf + "I kg of " +
                    TextBox1.Text.ToUpper + _
```

```
                          "produces " + Format(molCH4, "##.##")
+
                   " kg CH4 and " + Format(molCO2, "##.##") +
                   " kg CO2."
        End Sub




'****************************************************************
*******'*******************************************************
*************
    'Reads the chemical formula entered in TextBox1, goes
through it
    'letter-by-letter.
    'if the letter is 'C', it adds one to variable 'a', if it
is 'H', it adds to
    ''b', and so on. If there is a number after the letter,
e.g, 'H3' in
    'CH3COOH', it adds the value accordingly, in the above
example it
    'adds 3 to variable 'c', and so on...
'****************************************************************
*******'*******************************************************
*************
    Sub readChemicalFormula()
        Dim i As Integer = 0
        Dim formula As String = TextBox1.Text.ToUpper + " "
        a = 0 : b = 0 : c = 0 : d = 0
        While i < formula.Length - 1
            Select Case formula.Chars(i)
                Case "C"
                    If Val(formula.Chars(i + 1)) > 0 Then
                        a = a + Val(formula.Chars(i + 1))
                        i += 1
                    Else : a += 1
                    End If
                Case "H"
                    If Val(formula.Chars(i + 1)) > 0 Then
                        b = b + Val(formula.Chars(i + 1))
                        i += 1
                    Else : b += 1
                    End If
                Case "O"
                    If Val(formula.Chars(i + 1)) > 0 Then
                        c = c + Val(formula.Chars(i + 1))
                        i += 1
```

```
                    Else : c += 1
                    End If
              Case "N"
                    If Val(formula.Chars(i + 1)) > 0 Then
                        d = d + Val(formula.Chars(i + 1))
                        i += 1
                    Else : d += 1
                    End If
          End Select
          i += 1
      End While
  End Sub


    Private Sub Button1_Click(ByVal sender As System.Object,
            ByVal e As System.EventArgs) Handles
Button1.Click
        calculateResults()
    End Sub
End Class
```

In the beginning of composting process, mesophilic[12] microorganisms are frequent with most of occurring biochemical reactions being attributed to them. The increase in these organisms, after about a week, increases the temperature of compost which limits growth of these organisms to be replaced by thermophilic[13] organisms. When the temperature drops, it usually means that the compost needs to be aerated, or watered, or that composting is complete. It is desirable to work at a temperature between 60 to 75° C for complete digestion.

A critical variable in composting is the moisture content. If the mixture is too dry, the microorganisms cannot survive, and composting stops. If there is too much water, the oxygen from the air is not able to penetrate to where the microorganisms are, and the mixture becomes anaerobic. The right amount of moisture, whether

---

[12]   Mesophiles live in medium temperature ranging from 25 to 45oC.
[13]   Thermophiles are microorganisms that live at high temperatures, above 45oC.

wastewater sludge or other sources of water, that needs to be added to the solids to achieve just the right moisture content can be calculated from a simple mass balance as presented in equation 7.2.

$$M_P = \frac{M_a\, x_a + 100\ x_s}{x_s + x_a} \qquad\qquad (7.2)$$

Where:

M_P = Moisture in mixed pile (heap) ready to begin process of composting, %.

M_a = Moisture in solids as in shredded and screened refuse, %.

x_a = Mass of solids, wet tons.

x_s = Mass of sludge or other source of water, ton. (This assumes that the solids content of the sludge is very low, a good assumption if waste activated sludge is used, which is commonly less than 1 percent solids).

Example (7.2)

a) A mixture of paper, newspapers and potentially composted materials of mass 6 tons, amount of moisture content in it is 5 percent. It is required to make a mixture for the process of composting of moisture content of 50 percent moisture. Find amount of water or wastewater sludge to be added to the solids of this MSW to obtain desired concentration of moisture content of the heap to start the process of composting.

b) Write a computer program to determine the amount of water or wastewater sludge to be added to the solids of a MSW to obtain desired concentration of moisture content of the heap to start the process of composting given mass of composted materials of the MSW, amount of moisture content in it, and mixture moisture content for the process of composting.

c) Verify your program by solving example 7.2.

Solution

1) Given: Data: xa = 6 tons, Ma = 5%, MP = 50%.
2) Use equation to find amount of water needed xs from wastewater:

$$M_P = \frac{M_a\, x_a + 100\, x_s}{x_s + x_a}$$

$$50 = \frac{(5 \times 6) + 100\, x_s}{x_s + 6}$$

Then, xs = 5.4 tons from water or from wastewater sludge.

Program 7.2 Algorithm:
1. **Inputs:** xa (tons), Ma (%), MP (%)
2. **Calculations:** See Equation (7.2)
3. **Output:** Amount of water/wastewater sludge needed (xs, tons)

Program 7.2 Listing: amount of water or wastewater sludge to be added to solids of a MSW to obtain desired moisture of the heap to start composting

```
'*************************************************************
*******
'Program 7.2:Calculates amount of water or wastewater sludge to
be
'added to the solids of a MSW to obtain desired concentration
of
'moisture content.
'*************************************************************
*******
Public Class Form1
    Dim MP, Ma, Xa, Xs As Double

    Private Sub Form1_Load(ByVal sender As System.Object,
            ByVal e As System.EventArgs) Handles MyBase.Load
        Me.Text = "Program 7.2:"
        Me.FormBorderStyle =
            Windows.Forms.FormBorderStyle.FixedSingle
        Me.MaximizeBox = False

        Label1.Text = "The formula:" + vbCrLf
        Label1.Text += "MP = ((Ma * Xa) + (100 * Xs)) / (Xs +
Xa)"
```

131

```
        Label2.Text = "Composted materials mass, Xa (ton):"
        Label3.Text = "amount of moisture content in it, Ma
(%):"
        Label4.Text = "Moisture content, MP (%):"
        Label5.Text = ""
        Button1.Text = "&Calculate"
    End Sub

    Sub calculateResults()
        Xa = Val(TextBox1.Text)
        Ma = Val(TextBox2.Text)
        MP = Val(TextBox3.Text)

        'Formula: "MP = ((Ma * Xa) + (100 * Xs)) / (Xs + Xa)"
        Xs = ((MP * Xa) - (Ma * Xa)) / (100 - MP)
        Label5.Text = "xs = " + Format(Xs, "###.#") +
            " tons from water or from wastewater sludge."
    End Sub

    Private Sub Button1_Click(ByVal sender As System.Object,
            ByVal e As System.EventArgs) Handles
Button1.Click
        calculateResults()
    End Sub
End Class
```

Estimation of carbon and nitrogen levels and the C/N ratio is based on mass balances. If two materials such as shredded refuse and sewage sludge are mixed, the carbon of the mixture is calculated as shown in equation 7.3.

$$C_p = \frac{c_r\, X_r + c_s\, X_s}{X_r + X_s} \qquad (7.3)$$

Where
C_p = Carbon concentration in the mixture prior to composting, as percent of total wet mass of mixture.
C_r = Carbon concentration in the refuse, as percent of total wet refuse mass.

C_s = Carbon concentration in the sludge, as percent of total wet sludge mass.

X_s = Total mass of sludge, wet tons per day.

X_r = Total mass of refuse, wet tons per day.

7.2 Oxygen necessary for oxidizing hydrocarbons

MSW and refuse can be burned as is, or it can be processed to improve its heat value and to make it easier to handle in a combustor. Processed refuse also can be combined with other fuels (such as coal) and co-fired in a heat recovery combustor or can be used to provide electrical power for the community. Amount of oxygen necessary to oxidize some hydrocarbon is known as stoichiometric oxygen. Usually, refuse is not burned using air as the source of oxygen. Since air contains 23.15% oxygen by weight, then stoichiometric air required can be determined from equation 7.4.

Stoichiometric air = stoichiometric oxygen/ 0.2315   (7.4)

Example (7.3)

a) Calculate the stoichiometric oxygen and stoichiometric air required for the combustion of methane gas.

b) Write a computer program to calculate the stoichiometric oxygen and stoichiometric air required for the combustion of a certain organic gas.

c) Verify your program by solving example 7.3.

Solution

1) The equation for combustion of methane is
   $$CH_4 + 2O_2 = CO_2 + 2H_2O$$

2) That is, it takes 16 grams of methane (12 + 4) to react with 2x2x16 = 64 moles of oxygen. Thus, stoichiometric oxygen required for combustion of methane is 64/16 = 4 g O_2/g CH_4

3) Stoichiometric air requirement = 4/0.2315 = 17.3 g air/g methane.

Program 7.3 Algorithm: stoichiometric oxygen and stoichiometric air for combustion of an organic gas

1. **Inputs:** Equation of methane combustion
2. **Calculations:** See Equation (7.4)
3. **Output:** Stoichiometric air requirement

Program 7.3 Listing:

```
'***********************************************************
'Program 7.3: Calculates stoichiometric oxygen and air
'required for the combustion of a gas
'***********************************************************
Imports System.Math

Public Class Form1
    Dim a, b, c, d As Double
    Dim H2O, GAS, CO2, O2 As Double
    Dim outputStr(4) As String

    Private Sub Form1_Load(ByVal sender As System.Object,
            ByVal e As System.EventArgs) Handles MyBase.Load
        Me.Text = "Program 6.3:"
        Me.FormBorderStyle =
            Windows.Forms.FormBorderStyle.FixedSingle
        Me.MaximizeBox = False

        Label1.Text = "Enter the equation for combustion of the
gas:"
        Label2.Text = ""
        CheckBox1.Checked = False
        CheckBox1.Text = "Use Example 7.3, Combustion of
maethan
                         'CH4 + 2O2 = CO2 + 2H2O'"
        Button1.Text = "&Calculate"
    End Sub

    Sub calculateResults()
        readChemicalFormula()

        Dim result, result2 As Double
        result = O2 / GAS
        result2 = result / 0.2315
```

134

```
            Label2.Text = "It takes " + GAS.ToString + " grams of
("
                    + outputStr(1) + ") to react with"
            Label2.Text += O2.ToString + " moles of oxygen"
            Label2.Text += vbCrLf + "Stoichiometric oxygen required
for
                    combustion is " + Format(result, "###.#")
            Label2.Text += vbCrLf + "Stoichiometric air requirement
= " +
                    result.ToString + "/0.2315 = " _
                        + Format(result2, "###.#") + " g air/g
gas."
        End Sub



'*************************************************************
*******'*****************************************************
*************
    'Reads the chemical formula entered in TextBox1, goes
through it
    'letter-by-letter.
    'it divides the formula into four components, two on left
side of the
    'equal "=" sign
    'and two on the right side. it then calculates the weight
of each
    'one.
    'Sample formula: CH4 + 2O2 = CO2 + 2H2O
'*************************************************************
*******'*****************************************************
*************
    Sub readChemicalFormula()
        Dim i As Integer = 0
        Dim i2 As Integer = 0
        Dim formula As String = TextBox1.Text.ToUpper + " "
        Dim f(4) As String
        Dim r(4) As Double
        Dim x, multiplier As Double

        GAS = 0 : O2 = 0 : CO2 = 0 : H2O = 0
        'break down the formula into four parts
        i = formula.IndexOf("+")
        f(1) = formula.Substring(0, i)
        i2 = formula.IndexOf("=")
        f(2) = formula.Substring(i + 1, i2 - i - 1)
        i = formula.IndexOf("+", i2)
```

135

```
        f(3) = formula.Substring(i2 + 1, i - i2 - 1)
        f(4) = formula.Substring(i + 1, formula.Length - i - 1)

        For j = 1 To 4
            'run each part of the formula in turn, calculating
its weight
            i = 0
            multiplier = 0
            r(j) = 0
            f(j) = " " + f(j) + " "
            While i < f(j).Length - 1
                If multiplier = 0 And IsNumeric(f(j).Chars(i))
Then
                    'the first number in the formula is the
multiplier, e.g.
                    '2H2O
                        multiplier = Val(f(j).Chars(i))
                Else
                    Select Case f(j).Chars(i)
                        Case "C"
                            'if the next char is a number, e.g.
C2, then multiply
                    'this by the weight of carbon, then multiply
by the
                    'multiplier that was read early on if it is >
0.
                            If Val(f(j).Chars(i + 1)) > 0 Then
                                x = Val(f(j).Chars(i + 1)) * 12
'weight of Carbon
                                If multiplier > 0 Then
                                    r(j) = r(j) + x *
multiplier 'multiplier, e.g. 2*C
                                Else
                                    r(j) = r(j) + x
                                End If
                                i += 1
                            Else
                                If multiplier > 0 Then
                                    r(j) = r(j) + 12 *
multiplier 'multiplier, e.g. 2*C
                                Else : r(j) += 12
                                End If
                            End If
                        Case "H"
                            'if the next char is a number, e.g.
H2O, then
```

136

```
                        'multiply this by weight of hydrogen, then
multiply
                        'by the multiplier that was read early on if
it is > 0.
                            If Val(f(j).Chars(i + 1)) > 0 Then
                                x = Val(f(j).Chars(i + 1))
'weight of Hydrogen
                                If multiplier > 0 Then
                                    r(j) = r(j) + x *
multiplier 'multiplier, e.g. 2*H
                                Else
                                    r(j) = r(j) + x
                                End If
                                i += 1
                            Else
                                If multiplier > 0 Then
                                    r(j) = r(j) + multiplier
'multiplier, e.g. 2*H
                                Else : r(j) += 1
                                End If
                            End If
                        Case "O"
                            'if the next char is a number, e.g.
O2, then multiply
                            'this by weight of oxygen, then multiply by
the
                            'multiplier that was read early on if it is >
0.
                            If Val(f(j).Chars(i + 1)) > 0 Then
                                x = Val(f(j).Chars(i + 1)) * 16
'weight of Oxygen
                                If multiplier > 0 Then
                                    r(j) = r(j) + x *
multiplier 'multiplier, e.g. 2*O2
                                Else
                                    r(j) = r(j) + x
                                End If
                                i += 1
                            Else
                                If multiplier > 0 Then
                                    r(j) = r(j) + 16 *
multiplier
                        'multiplier, e.g. 2*O2
                                Else : r(j) += 16
                                End If
                            End If
```

```
                    End Select
                End If
                i += 1
            End While
        Next


        'make sure the right- and left-side of the equation are
balanced.
        If (r(1) + r(2)) <> (r(3) + r(4)) Then
            MsgBox("Equation is not balanced! Review and try
again.",
                        vbOKOnly, "Equation Error")
            Exit Sub
        End If


        'the four parts are not arranged.. arrange so that the
gas result
        'is the first, the oxygen is second, the CO2 is third,
and the
        'water is fourth.
        For i = 1 To 4
            If f(i).Contains("CO2") Then
                CO2 = r(i)
                outputStr(3) = f(i).Trim
            ElseIf f(i).Contains("O2") Then
                O2 = r(i)
                outputStr(2) = f(i).Trim
            ElseIf f(i).Contains("H2O") Then
                H2O = r(i)
                outputStr(4) = f(i).Trim
            Else
                GAS = r(i)
                outputStr(1) = f(i).Trim
            End If
        Next
    End Sub


    Private Sub Button1_Click(ByVal sender As System.Object,
            ByVal e As System.EventArgs) Handles
Button1.Click
        calculateResults()
    End Sub


    Private Sub CheckBox1_CheckedChanged(ByVal sender As
            System.Object, ByVal e As System.EventArgs)
            Handles CheckBox1.CheckedChanged
```

```
      If CheckBox1.Checked Then
          TextBox1.Text = "CH4 + 2O2 = CO2 + 2H2O"
      Else
          TextBox1.Text = ""
      End If
   End Sub
End Class
```

7.3 Energy balance

In a power plant, water is heated to steam in a boiler. Steam is used to turn a turbine, which drives a generator. This process can be simplified to a simple energy balance where energy in has to equal energy out (energy wasted in the conversion + useful energy) plus energy accumulated in box (energy changed in form) as presented in equation 7.5.

[Rate of energy accumulated] = [Rate of energy in] - [Rate of energy out] + [[Rate of energy produced] – [Rate of energy consumed]

$$(7.5)$$

Energy systems in steady state are defined as no change occurring over time. As such, there cannot be a continuous accumulation of energy, or if some of the energy out is useful and the rest is wasted and the equation 7.5 becomes 7.6.

[Rate of energy in] = [Rate of energy out]   (7.6)

[Rate of energy IN] = [Rate of energy used] + [Rate of energy wasted]   (7.7)

Efficiency (E %) of process can be calculated as indicted in equation 7.8.

$$E = \frac{\text{Energy used}}{\text{Energy in}} \times 100$$

$$(7.8)$$

A thermal balance on a large combustion unit is difficult because much of the heat cannot be accurately measured. Assuming recovery of heat as steam in a combustor, input heat to a black box (see figure 7.1) is from heat value in the fuel and heat in the water entering the water-wall pipes. The output is the sensible heat in the stack gases, the latent heat of water, the heat in the ashes, the heat in the steam, and the heat lost due to radiation. If the process is in a steady state, the equation of thermal balance can be as indicated in equation 7.9.

[Rate of heat accumulated] = [Rate of heat in the fuel] + [Rate of heat in the water] -[Rate of heat out in the stack gases] - [Rate of heat out in the stem] - [Rate of heat out as latent heat of vaporization] – [Rate of heat out in the ash] - [Rate of heat loss due to radiation]   (7.9)

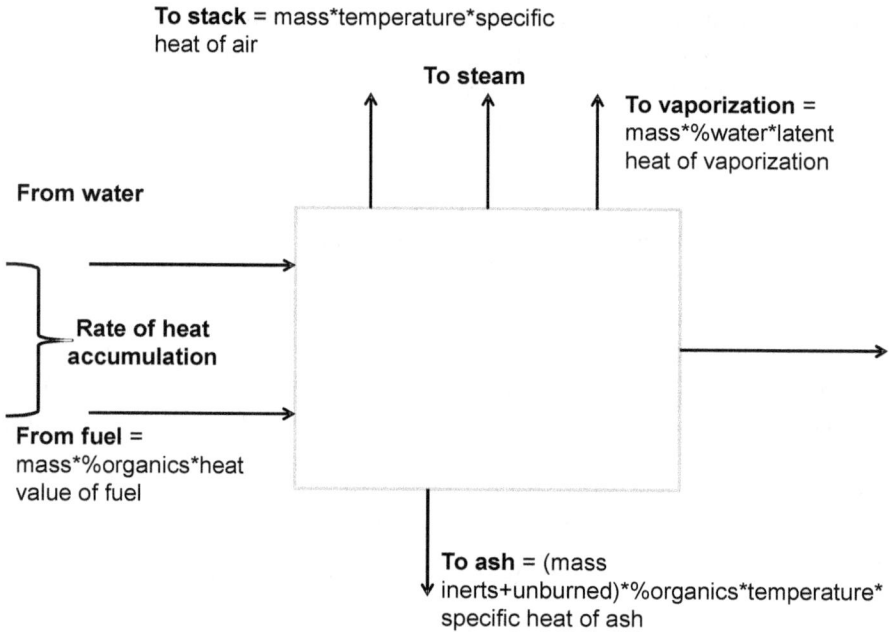

Figure 6.1: Energy flow in a combustor.

Example (7.4)

a) A refractory combustion unit lined with no water wall and no heat recovery is burning refuse-derived fuel (RDF) consisting of 80% organics, 12% water, and 8% inorganics or inerts at a rate of 19,920 kg/d. Determine the temperature of the stack gases, assuming the following:

140

- Heat value of the fuel = 19,000 kJ/kg on a moisture-free basis.
- Air flow = 8,300 kg/h and that the under- and over fire air contributes negligible heat.
- 7% of the heat input is lost due to radiation.
- 15% of the fuel remains un-combusted in the ash.
- ash exits the combustion chamber at a temperature = 750°C.
- Specific heat of ash = 0.837 kJ/kg/°C..
- Specific heat of air is 1.0 kJ/kg/°C.
- Latent heat of vaporization = 2575 kJ/kg.

b) Write a computer program to calculate the temperature of the stack gases for a refractory combustion unit lined with no water wall and no heat recovery burning a refuse-derived fuel (RDF) given: RDF percent organics, water, and inorganics or inerts, daily mass rate, heat value of the fuel on a moisture-free basis, air flow, percent heat input lost due to radiation, percent fuel remaining un-combusted in the ash, temperature of ash exiting combustion chamber, specific heat of ash, specific heat of air and latent heat of vaporization.

d) Verify your program by solving example 7.4.

Solution

1) Given: RDF 80% organics, 12% water, 8% inorganics or inerts , RDF rate = 19,920 kg/d = 19920/24 = 830 kg/hr, Heat value of the fuel = 19,000 kJ/kg, Air flow = 8,300 kg/hr, Heat input is lost due to radiation = 7%, Fuel remaining uncombusted in the ash = 15%, Ash temperature = 750°C, Specific heat of ash = 0.837 kJ/kg/°C, Specific heat of air is 1.0 kJ/kg/°C, Latent heat of vaporization = 2575 kJ/kg.

2) Draw balance box as follows:

141

To stack = heat of combustion – heat losses

To stack = mass*temperature*specific heat of air

= 12,616,000 – 104,206 –
883,120 – 256470

= 11,372,204 kJ/hr

To vaporization =
mass*%water*latent
heat of vaporization

= 830kg/h*0.12*2575
= 256,470 kJ/hr

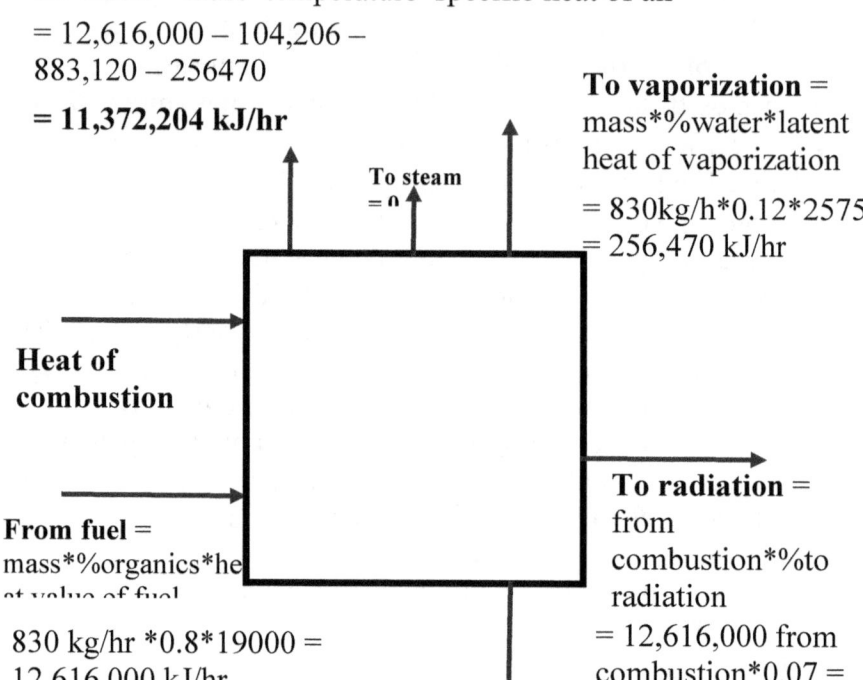

To steam
= 0

**Heat of
combustion**

From fuel =
mass*%organics*he
~~at value of fuel~~

830 kg/hr *0.8*19000 =
12,616,000 kJ/hr

To radiation =
from
combustion*%to
radiation

= 12,616,000 from
combustion*0.07 =
883,120 kJ/hr

To ash = (mass inerts+unburned)
*%organics*temperature*specific heat of

= (830*0.08 inerts + 830*0.8 organics
*0.15uncombusted)*750*0.837 = 104,206.5 kJ/hr

Energy flow in a combustor

Heat to stack = mass*temperature*specific heat of = 8300
kg/hr*T*1kJ/kg°C = 11372204
Temperature of stack gases = 1370 °C = (1370*9/5) +32 = 2499 °F

Program 7.4 Algorithm: temperature of stack gases for a refractory combustion unit burning a refuse-derived fuel

1. **Inputs:** RDF composition (%), RDF rate (kg/hr), Heat value of fuel (kJ/kg), Air flow (kg/hr), Heat lost to radiation (%), Uncombusted fuel remaining in ash (%), Air temp. (C), Specific heat of ash (kJ/kg/C) , Specific heat of air (kJ/kg/C), latent heat of vaporization (kJ/kg)
2. **Calculations:** See Equations (7.5), (7.6), (7.7), (7.8), (7.9), and Figure (7.1)
3. **Output:** The temperature of stack gases

Program 7.4 Listing:

```
'**************************************************************
'Program 7.4:calculates temperature of the stack gases
'**************************************************************
Public Class Form1
    Dim heatValue, airFlow, percRad, percAsh, exitTemp,
AshSpHeat,
            AirSpHeat, latentHeat As Double
    Dim RDFRate, percOrg, percWater, percInerts As Double
    Dim toStack, toVapor, toRad, toAsh, fromFuel As Double

    Private Sub Form1_Load(ByVal sender As System.Object,
            ByVal e As System.EventArgs) Handles MyBase.Load
        Me.Text = "Program 7.4:calculates temperature of the
stack
            gases"
        Me.FormBorderStyle =
            Windows.Forms.FormBorderStyle.FixedSingle
        Me.MaximizeBox = False

        Label1.Text = "Heat value of the fuel (kJ/kg):"
        Label2.Text = "Air flow (kg/hr):"
        Label3.Text = "Percentage of heat input lost due to
radiation
            (%):"
        Label4.Text = "Percentage of the remaining un-combusted
in
            the ash (%):"
```

143

```
        Label5.Text = "Ash exit temperature from combustion
chamber
              (C):"
        Label6.Text = "Specific heat of ash (C):"
        Label7.Text = "Specific heat of air (C):"
        Label8.Text = "Latent heat of vaporization (kJ/kg):"

        Label9.Text = "Refuse-derived fuel (RDF) rate (kg/d):"
        Label10.Text = "Percentage of organics (%):"
        Label11.Text = "Percentage of water (%):"
        Label12.Text = "Percentage of inorganics/inerts (%):"
        Button1.Text = "&Calculate"
    End Sub

    Sub calculateResults()
        heatValue = Val(TextBox1.Text)
        airFlow = Val(TextBox2.Text)
        percRad = Val(TextBox3.Text)
        percAsh = Val(TextBox4.Text)
        exitTemp = Val(TextBox5.Text)
        AshSpHeat = Val(TextBox6.Text)
        AirSpHeat = Val(TextBox7.Text)
        latentHeat = Val(TextBox8.Text)
        RDFRate = Val(TextBox9.Text)
        percOrg = Val(TextBox10.Text)
        percWater = Val(TextBox11.Text)
        percInerts = Val(TextBox12.Text)

        RDFRate = RDFRate / 24  'convert to kg/hr
        If percOrg > 1 Then percOrg /= 100 'if percentage is
input as xx
                                %, convert to 0.xx%
        If percWater > 1 Then percWater /= 100 'if percentage
is input
                                as xx%, convert to 0.xx%
        If percInerts > 1 Then percInerts /= 100 'if percentage
is input
                                as xx%, convert to 0.xx%
        If percRad > 1 Then percRad /= 100 'if percentage is
input as xx
                                %, convert to 0.xx%
        If percAsh > 1 Then percAsh /= 100 'if percentage is
input as xx
                                %, convert to 0.xx%

        fromFuel = RDFRate * percOrg * heatValue
```

```
            toVapor = RDFRate * percWater * latentHeat
            toRad = fromFuel * percRad
            toAsh = ((RDFRate * percInerts) + (RDFRate * percOrg *
                percAsh)) * exitTemp * AshSpHeat
            toStack = fromFuel - toVapor - toRad - toAsh

            drawStack()
        End Sub

    Sub drawStack()
        Dim g As Graphics
        g = PictureBox1.CreateGraphics
        g.Clear(Color.White)

        'draw stack box in the middle of the area
        g.DrawRectangle(Pens.Black, CInt((PictureBox1.Width /
2) - 50),_
                        CInt((PictureBox1.Height / 2) - 35),
100, 70)
        'draw the 7 arrows, 2 into and 5 out of the stack
        Dim x = CInt((PictureBox1.Width / 2) - 50)
        Dim y = CInt((PictureBox1.Height / 2) - 35)
        drawArrow(g, x - 30, y + 20, x, y + 20)
        drawArrow(g, x - 30, y + 50, x, y + 50)
        drawArrow(g, x + 20, y, x + 20, y - 20)
        drawArrow(g, x + 40, y, x + 40, y - 10)
        drawArrow(g, x + 60, y, x + 60, y - 30)
        drawArrow(g, x + 60, y + 70, x + 60, y + 100)
        drawArrow(g, x + 100, y + 40, x + 130, y + 40)

        'output the results
        Dim s As String
        Dim f As Font = New Font(FontFamily.GenericSansSerif,
8)
        Dim f2 As Font = New Font(f, FontStyle.Bold Or
                        FontStyle.Underline)
        s = "To stack = heat of combustion " + vbCrLf + "- heat
losses"
                        + vbCrLf + _
                        "= " + Format(toStack, "n") + "
kJ/hr"
        g.DrawString(s, f, Brushes.Black, New Point(10, y -
50))
        s = "To vaporization = mass*%water" + vbCrLf + "*latent
heat
                        of vaporization" + vbCrLf + _
```

145

```
                        "= " + Format(toVapor, "n") + "
kJ/hr"
        g.DrawString(s, f, Brushes.Black, New Point(x + 65, y -
50))
        s = "From fuel = mass" + vbCrLf + "*%organics*heat " +
vbCrLf
                    + "value of fuel" + _
                        vbCrLf + "= " + Format(fromFuel, "n")
+
                    " kJ/hr"
        g.DrawString(s, f, Brushes.Black, New Point(10, y +
30))
        s = "To radiation = from " + vbCrLf +
                "combustion*%to radiation" + vbCrLf + _
                    "= " + Format(toRad, "n") + " kJ/hr"
        g.DrawString(s, f, Brushes.Black, New Point(x + 110, y
+ 50))
        s = "To ash = (mass inerts+unburned)" + vbCrLf +
                "*%organics*temperature*specific" + _
                    vbCrLf + " heat of ash" + vbCrLf + "=
" +
                Format(toAsh, "n") + " kJ/hr"
        g.DrawString(s, f, Brushes.Black, New Point(x + 40, y +
110))

        s = "Energy flow in a combustor"
        g.DrawString(s, f2, Brushes.Black, New Point(x - 15, y
+ 160))
    End Sub


    Sub drawArrow(ByRef g As Graphics, ByVal x As Integer,
        ByVal y As Integer, ByVal x2 As Integer, ByVal y2 As
Integer)
            'Draw the arrow stem
            g.DrawLine(Pens.Black, x, y, x2, y2)
            'Draw the arrow head
            If x = x2 Then     'arrow is vertical
                If y2 > y Then     'arrow is facing down
                    g.DrawLine(Pens.Black, x2, y2, x2 - 4, y2 - 4)
                    g.DrawLine(Pens.Black, x2, y2, x2 + 4, y2 - 4)
                Else               'arrow is facing up
                    g.DrawLine(Pens.Black, x2, y2, x2 - 4, y2 + 4)
                    g.DrawLine(Pens.Black, x2, y2, x2 + 4, y2 + 4)
                End If
            Else               'arrow is horizontal
```

```
            If x > x2 Then     'arrow is facing left
                g.DrawLine(Pens.Black, x2, y2, x2 + 4, y2 - 4)
                g.DrawLine(Pens.Black, x2, y2, x2 + 4, y2 + 4)
            Else                'arrow is facing right
                g.DrawLine(Pens.Black, x2, y2, x2 - 4, y2 - 4)
                g.DrawLine(Pens.Black, x2, y2, x2 - 4, y2 + 4)
            End If
        End If
    End Sub

    Private Sub Button1_Click(ByVal sender As System.Object,
            ByVal e As System.EventArgs) Handles
Button1.Click
        calculateResults()
    End Sub
End Class
```

Exercise (7.1)

1) The three components of MSW of greatest interest in the bioconversion processes are: garbage (food waste), paper products, and yard wastes. What are the main factors that affect variation of garbage fraction of refuse? (B.Sc., UoD, 2012)

2) Theoretically, the combustion of refuse produced by a community is sufficient to provide about 20% of the electrical power needs for that community. Discuss this statement. (B.Sc., UoD, 2012)

Exercise (7.2)

1) Write down a balanced equation for the anaerobic decomposition of glucose. Estimate amount and volume produced (at STP)[14] of CO_2 and CH_4 during the anaerobic decomposition of glucose.

2) Estimate production of CO_2 and CH_4 during the anaerobic decomposition of municipal solid waste, MSW using the

---

[14] STP is used for expression of the properties and processes of ideal gases. The standard temperature is the freezing point of water and the standard pressure is one standard atmosphere. Standard temperature: 0°C = 273.15 K; Standard pressure = 1 atmosphere = 760 mmHg = 101.3 kPa ; Standard volume of 1 mole of an ideal gas at STP: 22.4 liters

147

chemical composition approximation of organic fraction of refuse as described by $C_{99}H_{149}O_{59}N$.

3) 12 tons of a mixture of paper and other compostable materials has a moisture content of 8%. The intent is to make a mixture for composting of 60% moisture. How many tons of water or sludge must be added to the solids to achieve this moisture concentration in the compost pile?

4) Methane and carbon dioxide generation by anaerobic digestion can be calculated using the following equation: (B.Sc., UoD, 2012)

$$C_aH_bO_cN_d + \left(\frac{4a-b-2c+3d}{4}\right)H_2O$$

$$\rightarrow \left(\frac{4a+b-2c-3d}{8}\right)CH_4 + \left(\frac{4a-b+2c+3d}{8}\right)CO_2 + dN\ H_3$$

Given the chemical composition of ethanol C_2H_5OH, propionic acid CH_3CH_2COOH and butyric acid $C_4H_8O_2$

  a) Write down the chemical reaction equations for the production of CO_2 and CH_4 during any anaerobic decomposition of the stated compounds.

  b) Compute amount of methane and carbon dioxide to be produced from anaerobic digestion of one kilogram of each compound.

  c) Determine total volume, in liters, of gases produced at STP.

5) A coal-fired power plant uses 800 Mg^{15} of coal per day. The energy value of the coal is 28,000 kJ/kg (kiloJoules/kilogram). The plant produces 3.2×10^6 kWh^{16} of electricity each day. What is the efficiency of the power plant?

6) A combustion unit is burning refuse-derived fuel (RDF) consisting of 75% organics, 15% water, and 10% inorganics

---

[15]   Megagrams, or 1000 kg, commonly called a metric tonne.
[16]   1 kWh = 3.6x10⁶ Joule

(inerts) at a rate of 1000 kg/h. Compute the temperature of the stack gases assuming the following:

a) Heat value of the fuel = 15,000 kJ/kg on a moisture-free basis.
b) Unit is refractory lined with no water wall and no heat recovery andd under- and overfire air contributes negligible heat.
c) Air flow = 8,000 kg/h.
d) 8% of the heat input is lost due to radiation
e) 10% of the fuel remains uncombusted in the ash,
f) Ash exits the combustion chamber at a temperature of 600°C.
g) specific heat of ash is 0.837 kJ/kg/°C,
h) Specific heat of air is 1.0 kJ/kg/°C.
i) Latent heat of vaporization = 2258 kJ/kg.

7) A refractory combustion unit lined with no water wall and no heat recovery is burning refuse-derived fuel (RDF) consisting of 85% organics, 10% water, & 5% inorganics or inerts at a rate of 950 kg/hr. Determine the temperature, T, at which ash exits the combustion chamber (in both °C and °F), assuming the following: (B.Sc., UoD, 2012)

Heat value of the fuel = 19,000 kJ/kg on a moisture-free basis.
Air flow = 9,500 kg/h & that the under- & overfire air contributes negligible heat.
5% of the heat input is lost due to radiation
15% of the fuel remains uncombusted in the ash
Specific heat of ash = 0.837 kJ/kg/°C,
Specific heat of air is 1.0 kJ/kg/°C.
Latent heat of vaporization = 2575 kJ/kg.
Temperature of the stack gases = 1500°C

Chapter Eight

Financing Solid Waste Facilities

8.1 Background

Solid waste may be defined as garbage, refuse and other solid material derived from any agricultural, commercial, consumer or industrial operation or activity if it is both: used material or residual material, and reasonably expected to be introduced into a qualified[17] solid waste disposal process within a reasonable time after such purchase or acquisition.

Solid waste financing or funding concerns revenues and costs which vary with specifics of the solid waste system, ownership and contractual arrangements and complexity of financial system. Revenue and profit for solid waste operations may be received from: sale of services and goods, garbage bill paid by home or business, tipping fees at disposal site, sale of recyclables and sale of products such as landfill gas or electricity from waste-to-energy plant. The

---

[17] A qualified solid waste disposal process may employ any biological, engineering, industrial or technological method. Eligible types of solid waste disposal processes include a final disposal process, an energy conversion process and a recycling process. A final disposal process is either the placement of solid waste in a landfill, the incineration of solid waste without capturing any useful energy, or the containment of solid waste with a reasonable expectation that the containment will continue indefinitely and that the solid waste has no current or future beneficial use. Energy conversion process encompasses a thermal, chemical or other process that is applied to solid waste to create and capture synthesis gas, heat, hot water, steam or other useful energy. The energy conversion process ends before any transfer or distribution of synthesis gas, heat, hot water, steam or other useful energy. Recycling process regards a process reconstituting, transforming or otherwise processing solid waste into a useful   product (http://www.squiresanders.com/tax_exempt_financing_of_solid_waste_dispos al_facilities/).

initial cost of the facility (or its capital cost) is an important one-time investment that may be paid from budget of the municipality or agency, or proceeds of bank loans, or general obligation bonds, or revenue bonds etc...

8.2 Capital cost and capital recovery factor

The capital costs of competing facilities can be projected by determining the cost that the municipality or agency would incur if it were to pay interest on a loan of that amount and value. Computing the annual cost of a capital investment resembles computing the annual cost of a loan or mortgage on a building or land. The municipality or agency borrows the money from a moneylender or a financier or bank and then has to pay it back in a number of equal installments. If the municipality borrows (X) dollars and aims to pay back the loan in (n) number of installments at an interest rate of (i), each installment can be found as presented in equation 8.1.

$$Y = \left[\frac{i(1+i)^n}{(1+i)^n - 1} \right] X \qquad (8.1)$$

Where
Y = Installment cost.
i = Annual interest rate (enter interest (i) in decimal form i.e. as a fraction).
n = Number of installments.
X = Amount borrowed.

A capital recovery factor, CRF, is defined as shown in equation 8.2

$$CRF = \left[\frac{i(1+i)^n}{(1+i)^n - 1} \right] \qquad (8.2)$$

Equations 7.1 and 7.2 can be combined as revealed in equation 8.3.

$$Y = CRF*X \qquad (8.3)$$

151

Example 8.1

a) A municipality decides to purchase a refuse collection truck that has an expected life of 10 years for SAR[18]590,000. Cost of the truck is to be borrowed from the local bank and to be paid back in 10 annual payments. Determine the annual installments on this capital expense if the interest rate is 6.125%

b) Write a computer program to determine the annual installments on capital expense for a municipality that decided to purchase a refuse collection truck given trucks' expected life for a given amount, cost of the truck to be borrowed from the local bank, annual back payments, and the interest rate.

c) Verify your program by solving example 8.1.

Solution

1. Given: t = 10 yr., cost of truck, X = SAR 590,000, payments = 10, i = 6.125%.

2. From Table (7.2), the capital recovery factor, CRF for n years = 10 is 0.13667. or:

$$\left[CRF = \frac{i(1+i)^n}{(1+i)^n - 1} \right] = \left[\frac{0.06125(1+0.06125)^{10}}{(1+0.06125)^{10} - 1} \right] = 0.136674$$

3. The annual cost to the municipality would then be, Y = CRF * X = 0.13667 * 590,000 or Y = SAR 80,636.

4. That is, the municipality would have to pay SAR 80,636 each year for 10 years to pay back the bank loan on this truck.

5. It is to be noted that this truck does not cost 10 * 80,636 = SAR 800,636, because the Saudi riyals for each year are different and cannot be augmented and added.

Program 8.1 Algorithm: annual installments on capital expense for a municipality purchasing a refuse collection truck

1. **Inputs:** t (yr), X (SAR), No. of payments, i (%)

---

[18]   1 United States dollar, US$ ≈ 3.8 Saudi Arabia Riyal, SAR.

2. **Calculations:** CRF, See Equations (8.1), (8.2), and (8.3)
3. **Output:** The annual installments (Y, SAR)

Program 8.1 Listing:

```
'****************************************************************
*****
'Program 8.1: Calculates capital cost, capital recovery factor
'and annual installments
'****************************************************************
*****
Public Class Form1
    Dim exLife, price, backPay, intRate As Double

    Private Sub Form1_Load(ByVal sender As System.Object,
            ByVal e As System.EventArgs) Handles MyBase.Load
        Me.Text = "Program 7.1:"
        Me.FormBorderStyle =
            Windows.Forms.FormBorderStyle.FixedSingle
        Me.MaximizeBox = False

        Label1.Text = " Calculates capital cost, capital
recovery factor
                    and annual installments."

        Label2.Text = "Truck expected life (yr):"
        Label3.Text = "Price (SAR):"
        Label4.Text = "Annual back payments:"
        Label5.Text = "Interest rate (%):"

        Button1.Text = "&Calculate"
        Label6.Text = ""
    End Sub

    Sub calculateResults()
        exLife = Val(TextBox1.Text)
        price = Val(TextBox2.Text)
        backPay = Val(TextBox3.Text)
        intRate = Val(TextBox4.Text)

        intRate /= 100
        Dim CRF, aCost As Double
        CRF = ((intRate * ((1 + intRate) ^ exLife)) / (((1 +
intRate) ^
            exLife) - 1))
```

```
        aCost = CRF * price

        Label6.Text = "CRF for " + backPay.ToString + " years
is " +
                Format(CRF, "n")
        Label6.Text += vbCrLf + "The annual cost would be, Y =
" +
                Format(CRF, "n") + _
                    " * " + backPay.ToString + " = SAR "
+
                Format(aCost, "n")
    End Sub

    Private Sub Button1_Click(ByVal sender As System.Object,
            ByVal e As System.EventArgs) Handles
Button1.Click
        calculateResults()
    End Sub
End Class
```

8.3 Present worth value and present worth factor

The actual cost of a capital investment also may be estimated by evaluating the present worth value or the value on a given date of a payment made at other times. This concerns finding amount to be invested at the moment (present), (Y) dollars, at a certain interest rate (i) to have available (X) dollars every year for (n) number of years. The relationship can be figures as presented in equation 8.4.

$$Y = \left[\frac{(1+i)^n - 1}{i(1+i)^n}\right] X \qquad\qquad (8.4)$$

Where
Y = Amount that has to be invested.
i = Annual interest rate.
n = Number of years.
X = Amount available every year.

A present worth factor, PWF can be introduced as shown in equation 8.5

154

$$PWF = \left[\frac{(1+i)^n - 1}{i(1+i)^n} \right] \qquad (8.5)$$

Then, by combining equations 8.4 and 8.5, equation 8.6 is obtained.

$$Y = PWF*X \qquad (8.6)$$

Example 8.2

a) A town wants to invest money in a bank account drawing an interest rate of 6.125% so that it can withdraw SAR 80,636 every year for the next 10 years. Compute the amount that must be invested?

b) Write a computer program to determine the amount that must be invested by a town who wants to invest money in a bank account given drawing interest rate; yearly withdraw amount and number of withdrawal years.

c) Verify your program by solving example 8.2.

Solution

1. Given: $i = 6.125\%$, withdraw, $X = $ SAR 80,636, $n = 10$ yr.
2. From Table (7.2), the present worth factor (PWF) for $n = 10$ is 7.316.
3. Thus, the money required, $Y = PWF*X = 7.316$ x SAR 80,636 or $Y = $ SAR 590,000.

Program 8.2 Algorithm: amount to be invested by a town

1. **Inputs:** n (yr), X (SAR), No. of payments, i (%)
2. **Calculations:** See Equations (7.4), (7.5), and (7.6)
3. **Output:** The required money (Y, SAR)

Program 8.2 Listing:

```
'**************************************************************
*****
'Program 8.2: Calculates amount of investment needed
'for a number of years.
```
155

```vb
'*************************************************************
*****
Public Class Form1
    Dim i, X, n, Y As Double

    Private Sub Form1_Load(ByVal sender As System.Object,
            ByVal e As System.EventArgs) Handles MyBase.Load
        Me.Text = "Program 8.2:"
        Me.FormBorderStyle =
            Windows.Forms.FormBorderStyle.FixedSingle
        Me.MaximizeBox = False

        Label1.Text = " Calculates amount of investment needed
for a
                    number of years."

        Label2.Text = "Interest rate (%):"
        Label3.Text = "Yearly withdrawal needed (SAR):"
        Label4.Text = "Total number of years:"

        Button1.Text = "&Calculate"
        Label5.Text = ""
    End Sub

    Sub calculateResults()
        i = Val(TextBox1.Text)
        X = Val(TextBox2.Text)
        n = Val(TextBox3.Text)

        i /= 100
        Dim PWF As Double
        PWF = (((((1 + i) ^ n) - 1) / (i * ((1 + i) ^ n)))
        Y = PWF * X

        Label5.Text = "Present worth factor (PWF) for " +
n.ToString +
                    " years is " + Format(PWF, "n")
        Label5.Text += vbCrLf + "The money required, Y = " +
                    Format(PWF, "n") + _
                        " * " + Format(X, "n") + " = SAR " +
                    Format(Y, "n")
    End Sub

    Private Sub Button1_Click(ByVal sender As System.Object,
            ByVal e As System.EventArgs) Handles
Button1.Click
```

```
        calculateResults()
    End Sub
End Class
```

8.4 Sinking fund and sinking fund factor

Sinking fund may be defined as a fund established by a municipality or government agency or business for the purpose of reducing debt by repaying or purchasing outstanding loans and securities held against the entity. This indicates that a sinking fund would be a sum of money set up to collect a certain amount of money to pay for purchasing a certain commodity or paying the bill of a major work. This means that the municipality or agency is saving money by investing it so that at some later date it would have available some specified sum. An example of such a fund in solid waste engineering is for the case of landfill entity investing money during the active life of the landfill so that, when the landfill is full, sufficient funds would be available to place the required final landfill cover.

Equation 8.7 illustrates how to determine the funds (Y) necessary to be invested in an account that draws (i) percent interest so that at the end of (n) years the fund has (X) value in it.

$$Y = \left[\frac{i}{(1+i)^n - 1} \right] X \qquad (8.7)$$

A sinking fund factor, SFF may be introduced as shown in equation 8.8.

$$SFF = \left[\frac{i}{(1+i)^n - 1} \right] \qquad (8.8)$$

Example 8.3

a) A local solid waste enterprise aspires to have SAR 2 million available at the end of a 10 year period by investing annually into an account that gives an interest of 6.125%. Find the amount the enterprise has to invest annually?

157

b) Write a computer program to determine the amount a local solid waste enterprise has to invest annually a certain amount available at the end of a 10 year period by investing annually into an account given interest rate.

c) Verify your program by solving example 8.3.

Solution

1. Given: X = SAR 2,000,000, t = 10 yr., i = 6.125%
2. From Table (7.2), the sinking fund factor (SFF) at 10 years is 0.07452.
3. The required annual investment is therefore Y = SFF*X = 0.07452 x SAR 2,000,000 or Y = SAR 149,040.
4. Note that the value of money of 10*149,040 = SAR 1,490,040 is significantly less than SAR 2,000,000. This is because the investments during the early years are drawing interest and adding to the sum available.

Program 8.3 Algorithm: annual amount to be invested a local solid waste enterprise

1. **Inputs:** t (yr), X (SAR), i (%)
2. **Calculations:** SFF, See Equations (8.7), and (8.8)
3. **Output:** The required money (Y, SAR)

Program 8.3 Listing:

```
'*************************************************************
*****
'Program 8.3: Calculates necessary funds to reach a certain
value
'*************************************************************
*****
Public Class Form1
    Dim i, X, n, Y As Double

    Private Sub Form1_Load(ByVal sender As System.Object,
            ByVal e As System.EventArgs) Handles MyBase.Load
        Me.Text = "Program 8.3:"
        Me.FormBorderStyle =
            Windows.Forms.FormBorderStyle.FixedSingle
```

```
        Me.MaximizeBox = False

        Label1.Text = "Calculates necessary funds to reach a
certain
                        value"

        Label2.Text = "Interest rate (%):"
        Label3.Text = "Total number of years:"
        Label4.Text = "Target value at the end of investment
period
                        (SAR):"

        Button1.Text = "&Calculate"
        Label5.Text = ""
    End Sub

    Sub calculateResults()
        i = Val(TextBox1.Text)
        n = Val(TextBox2.Text)
        X = Val(TextBox3.Text)

        i /= 100
        Dim SFF As Double
        SFF = i / (((1 + i) ^ n) - 1)
        Y = SFF * X

        Label5.Text = "Sinking fund factor (SFF) at " +
n.ToString + "
                        years is " + Format(SFF, "n")
        Label5.Text += vbCrLf + "The required annual
investment, Y = "
                        + Format(SFF, "n") + _
                        " * " + Format(X, "n") + " = SAR " +
                        Format(Y, "n")
    End Sub

    Private Sub Button1_Click(ByVal sender As System.Object,
            ByVal e As System.EventArgs) Handles
Button1.Click
        calculateResults()
    End Sub
End Class
```

Table (8.1) gives a general summary of selected compounding factors.

Table (8.1) Summary of selected compounding factors.

Factor	Abbreviation	Equation	Use	Examples
Compound amount	CA	$\left[(1+i)^n\right]$		
Capital recovery factor	CRF	$\left[\dfrac{i(1+i)^n}{(1+i)^n-1}\right]$	Pay loan back in a number of equal installments. Converts a present value into a stream of equal annual payments over a specified time.	
Present worth factor	PWF	$\left[\dfrac{(1+i)^n-1}{i(1+i)^n}\right]$	how much to be invested right now	
sinking fund factor	SFF	$\left[\dfrac{i}{(1+i)^n-1}\right]$	saving money by investing it so that at some later date some specified sum would be available	When landfill owner must invest money during active life of landfill so that, when landfill is full, there are sufficient funds to place final cover.

The capital recovery factor, present worth factor and sinking fund factors need not be computed, since they can be found in interest tables or are programmed into hand-held calculators or computer software. Table (8.2) shows these capital recovery factors for an interest rate of 6.125%.

160

Table (8.2) Capital recovery factors for an interest rate of 6.125%

Year	CRF	PWF	SFF
1	1.06125	0.942285041	1
2	0.546392511	1.83018614	0.485142511
3	0.374975336	2.666842064	0.313725336
4	0.289417974	3.455210425	0.228167974
5	0.238204237	4.198078139	0.176954237
6	0.204161994	4.898071274	0.142911994
7	0.179931702	5.557664333	0.118681702
8	0.161833535	6.179189007	0.100583535
9	0.147823104	6.764842409	0.086573104
10	0.136673733	7.31669485	0.075423733
11	0.127604778	7.836697149	0.066354778
12	0.120095776	8.326687537	0.058845776
13	0.113786379	8.788398151	0.052536379
14	0.10841917	9.223461155	0.04716917
15	0.103805354	9.633414516	0.042555354
16	0.099803313	10.01970744	0.038553313
17	0.096304735	10.38370548	0.035054735
18	0.093225356	10.72669538	0.031975356
19	0.090498641	11.04988964	0.029248641
20	0.088071346	11.35443076	0.026821346

Key:

$$CRF = \text{Capital Recovery Factor} = \left[\frac{i(1+i)^n}{(1+i)^n - 1} \right]$$

$$PWF = \text{Present Worth Factor} = \left[\frac{(1+i)^n - 1}{i(1+i)^n} \right]$$

$$SFF = \text{Sinking Fund Factor} = \left[\frac{i}{(1+i)^n - 1} \right]$$

8.5 Total cost

Total cost to the community is the sum of the annual payback of the capital costs (fixed costs) of the investments and the labor and raw materials costs and operating and maintenance costs (variable costs).

Example 8.4

a) A municipality wants to purchase a refuse collection truck that has an expected life of 10 years and costs SAR 590,000. The municipality preferred to pay back the loan in 10 annual installments at an interest rate of 6.125%. The annual operation cost of the truck (gas, oil, service and regular maintenance) amounts to SAR 80,000. How much will this truck cost the municipality every year?

b) Write a computer program to determine yearly cost a municipality ought to pay for a refuse collection truck given: trucks' expected life, and costs, annual installments for paying back loan, interest rate, and annual operation cost of the truck (gas, oil, service and regular maintenance) amounts.

c) Verify your program by solving example 8.4.

Solution

1. Given: n = 10 years, X = SAR 590,000, i = 6.125%. truck operating cost = SAR 80,000/yr
2. From Table, the capital recovery factor for n = 10 is 0.13667
3. so the annual cost of the capital investment is y = CRF*X = 0.13667 x SAR 590,000 or Y = SAR 80,636.
4. Total annual cost to the community = operating cost + annual investment = SAR 80,636 + SAR 80,000 or Total cost = SAR 160,,636.

Program 8.4 Algorithm: yearly cost to be paid by a municipality for a refuse collection truck

1. **Inputs:** n (yr), X (SAR), i (%), operating cost (SAR/yr)
2. **Calculations:** CRF (SAR), Y (SAR), See Tables (8.1), and (8.2)

3. Output: The annual cost (SAR)

Program 8.4 Listing:

```
'***************************************************************
*****
'Program 7.4: Calculates refuse collection truck's costs
'***************************************************************
*****
Public Class Form1
    Dim i, X, n, Y, cost As Double

    Private Sub Form1_Load(ByVal sender As System.Object,
            ByVal e As System.EventArgs) Handles MyBase.Load
        Me.Text = "Program 8.4:"
        Me.FormBorderStyle =
            Windows.Forms.FormBorderStyle.FixedSingle
        Me.MaximizeBox = False

        Label1.Text = "Calculates refuse collection truck's
costs"

        Label2.Text = "Expected life of collection truck:"
        Label3.Text = "Truck cost (SAR):"
        Label4.Text = "Annual installments interest rate (%):"
        Label5.Text = "Truck operating cost:"

        Button1.Text = "&Calculate"
        Label6.Text = ""
    End Sub

    Sub calculateResults()
        n = Val(TextBox1.Text)
        X = Val(TextBox2.Text)
        i = Val(TextBox3.Text)
        cost = Val(TextBox4.Text)

        i /= 100
        Dim CRF, total As Double
        CRF = (i * ((1 + i) ^ n)) / (((1 + i) ^ n) - 1)
        Y = CRF * X
        total = Y + cost

        Label6.Text = "Capital recovery factor (CRF) for " +
n.ToString +
                    " years is " + Format(CRF, "n")
```

163

```
        Label6.Text += vbCrLf + "The required annual
investment, Y = "
                        + Format(CRF, "n") + _
                          " * " + Format(X, "n") + " = SAR " +
                        Format(Y, "n")


        Label6.Text += vbCrLf + "The annual cost = " +
                    Format(cost, "n") + _
                    " + " + Format(Y, "n") + " = SAR " +
              Format(total, "n")
    End Sub

    Private Sub Button1_Click(ByVal sender As System.Object,
          ByVal e As System.EventArgs) Handles
Button1.Click
        calculateResults()
    End Sub
End Class
```

Exercise (8)

1. A municipality decides to purchase a refuse collection truck that has an expected life of 15 years for SAR 700,000. Cost of the truck is to be borrowed from the local bank and to be paid back in 15 annual payments. Determine the annual installments on this capital expense if the interest rate is 6.125%

2. A municipality decides to purchase a refuse collection truck that has an expected life of 20 years for SAR 980,000. Cost of the truck is to be borrowed from the local bank and to be paid back in 20 annual payments. Determine the annual installments on this capital expense if the interest rate is 6.125%

3. A municipality chooses to purchase a refuse collection truck that has an expected life of 15 years for the value of SAR 650,000. The cost of the truck is to be borrowed from the local bank and to be paid back in 10 annual payments. Given that the annual installments on this capital expense is SAR 67,500. Find the rate of interest.

4. A town wants to invest money in a bank account drawing an interest rate of 6.125% so that it can withdraw SAR 90,500

every year for the next 10 years. Compute the amount that must be invested?

5. A local solid waste enterprise aspires to have SAR 1.8 million available at the end of a 10 year period by investing annually into an account that gives an interest of 6.125%. Find the amount the enterprise has to invest annually?

6. A municipality wants to purchase a refuse collection truck that has an expected life of 15 years and costs SAR 700,000. The municipality preferred to pay back the loan in 15 annual installments at an interest rate of 6.125%. The annual operation cost of the truck (gas, oil, service and regular maintenance) amounts to SAR 72,000. How much will this truck cost the municipality every year?

Chapter Nine

General Questions

9.1 Matching questions

Rearrange group (I) with the corresponding relative ones of group (II) in the area allocated for the answer. (B.Sc., UoD, 2013)

Group (I)	Rearranged group (II)	Group (II)
Recyclables	**Newspapers**	Food waste
Refrigerators	**White waste**	Hospitals
Bulky refuse	**Furniture**	Refineries
Yard waste	**Green waste**	Chemical & Biological processes
Biomaterials	**Hospitals**	Rubble & remnants of Buildings
Sweeping	**Thrown by users**	Diverted refuse
Commercial waste	**Warehouse waste**	Rest and stability
Municipal waste	**Food waste**	ASTM Standard
Industrial refuse	**Refineries,**	Geotechnical engineering
Agriculture waste	**Livestock farms**	Generated by households
Hazardous waste	**Chemical & Biological processes**	Newspapers
Construction and demolition	**Rubble & remnants of Buildings**	Quartering and coning
Refuse	**Generated by households**	Production of gas
Dry weight moisture	**Geotechnical engineering**	Thrown by users
Waste sampling	**ASTM Standard**	Livestock farms
Representative samples	**Quartering and coning**	Green waste
Not collected waste	**Diverted refuse**	Furniture
Organic materials	**Production of gas**	Warehouse waste
Angle of repose	**Rest and stability**	White waste

166

9.2 Missing titles

Complete missing titles by using the following words and phrases (flooding. providers. odors. moisture. syringes) (B.Sc., UoD, 2013)

1) Garbage piles up on roads, streets and parks producing foul **odors.**
2) Bodies that deal with solid waste and garbage include: citizens on a daily basis, **providers** of waste collection services, scavengers and those working on re-use.
3) Accumulation of waste in drainage networks and waterways increases risk of **flooding** and contamination of water resources.
4) Types of hazardous solid waste include: linens, clothing, bandages & disposable, **syringes**, needles, surgical equipment & medical devices disposed off, food & contaminated waste and flammable materials.
5) Common contaminants of waste items include **moisture**, food, and dirt.

9.3 True-False questions

Indicate whether the following sentences are true (T) or false (F): (B.Sc., UoD, 2013)

1) Minimizing waste generated by the society & converting it to into a resource is the essence of the zero waste concept. (**T**)
2) Municipal solid waste defines a heterogeneous mixture of refuse, construction & demolition waste, leaves and bulky items. (**T**)
3) It is difficult to quantify hazardous waste due to lack of real statistics, & perhaps hiding it from producers. This calls for sudden visits and regular monitoring. (**T**)
4) Factors that affect quality & quantity of waste produced include: standards, laws & legislation in force, living conditions, urbanization in the region and social & economic factors. (**T**)
5) Low income areas generate less waste but with higher proportion of food. (**T**)
6) Properties of solid waste affect design of collection systems, treatment and disposal, operation, management and performance of units. (**T**)

167

References

1) Abdel-Magid, I. M., Hago, A. and Rowe, D. R., Modeling methods for environmental engineers, CRC Press/ Lewis Publishers, Boca Raton FL, 1995.

2) Abdel-Magid, I.M, Solid waste engineering and management, Sudan Academy Distributing and Publishing House, Scientific Books Series No. 1, Sudan Academy for Sciences, Khartoum, 2006 (In Arabic).

3) Anschutz, J., Ijgosse, J., and Scheinberg, A., Putting integrated sustainable waste management into practice. Using the ISWM assessment methodology, WASTE, Gouda, The Netherlands, 2004.

4) Blackman, W. C., Basic hazardous waste management, Lewis Publishers CRC Press LLC, Boca Raton, 2001.

5) Brown, R. P., Greenwood, J. H., Practical Guide to the Assessment of the Useful Life of Plastics, ERA Technology Ltd., Shropshire, SY, 2002.

6) de Bertoldi, M., Science of composting, Springer; 1996,

7) Chandler, A. J., Eighmy, T. T., Hartlen, J., Hjelmar, O., Kosson, D. S., Sawell, S. E., Vehlow, J., Municipal solid waste incinerator residues, the international ash working group, Studies in Environmental Science 67, Elsevier, Amsterdam, 1997.

8) Cheremisinoff, N. P., handbook of solid waste management and Waste minimization technologies, Burlington, MA, Elsevier Science, 2003.

9) Cheremisinoff, N. P., Consulting engineer handbook of solid waste management and waste minimization technologies, Butterworth-Heinemann.

10) CEHA, Solid waste management in some countries of the Eastern Mediterranean region, WHO, Eastern Mediterranean Regional Office, Regional Centre for Environmental Health Activities, Amman, Jordan, CEHA Document No. , Special studies, ss-4, 1995.

11) Davis, M. L. and Cornwell, D. A., Introduction to environmental engineering, McGraw-Hill Science, New York, 2006.

12) Dulac, N., The organic waste flow in integrated sustainable waste management, Waste, Gouda, The Netherlands, 2001.

13) Envirodyne, E., Beveridge and Diamond, P. C., Municipal solid waste management options Vol. 1-4, Illinois Department of Energy and Natural Resources, Office of Solid Waste and Renewable Resources, Spring field, IL, ILENR/RR- 89/06, 1989.

14) Franchetti, M. J., Solid waste analysis and minimization: a systems approach: The systems approach, McGraw-Hill Companies, Inc., New York, 2009.

15) Gören, S., Sanitary Landfill, Fatih University, Istanbul, 2004Henry, J. G. and Heinke, G. W., Environmental science and engineering, Prentice Hall, Englewood Cliffs. N. J., 1989.

16) Hibrawi, K., Engineering encyclopedia environmental treatment of solid waste, Saudi Aramco DeskTop Standards.

17) Hoornweg. D., Thomas, L. and Otten, L., Composting and its applicability in developing countries, Working Paper Series, 8, Published for the Urban Development Division, The World Bank, Washington, DC., 1999.

18) Johannessen, L. M. and Boyer, G., Observations of solid waste landfills in developing countries, Africa, Asia and Latin America, Urban Development Division, Waste Management Anchor Team, The World Bank, Washington, DC., 1999.

19) Kindlein, J., Dinkler, D. and Ahrens, H., Verification and application of coupled models for transport and reaction process in sanitary landfills, Proceedings Sardinia, Ninth International Waste Management and Landfill Symposium, S. Marghorita di Pula, Cagliari, Italy, 6 – 10 October 2003.

20) Kumar, E. S., Waste management, Intech, Olajnica, India, 2010.

21) Maczulak, A. E., Cleaning up the environment: hazardous waste technology, Anne Maczulak, New York NY, 2009.

22) McDougall, F.R., White, P.R., Franke, M. and Hindle, P., Integrated solid waste management: a life cycle inventory, Blackwell Science Ltd, Oxford, 2009.

23) Ojovan, M. I., Edi., Handbook of advanced radioactive waste conditioning technologies, Woodhead Publishing Ltd., 2011

24) Peavy, H. S., Rowe, D. R., Tchobanglous, G., Environmental engineering, McGraw-Hill Book Co., New York, 1985.

25) Perry, R. H., Green, D. W. and Maloney, J. O., Edi., Perry's Chemical Engineers' Handbook, McGraw-Hill Professional; 8 edition, 2007.

26) Popel, J. H., Storage, collection and transportation of domestic refuse, Delft University of Technology. Delft, 1971.

27) Proceedings Sardinia, Ninth International Waste Management and Landfill Symposium, S. Marghorita di Pula, Cagliari, Italy, 745 scientific papers, 6 – 10 October 2003

28) Rodic-wiersma, L., Introduction to solid waste management and engineering, Refresher course on solid waste management and engineering, organized by UNESCO-IHE Institute for water education, Delft, The Netherlands, 16 – 22 October 2005, Mombasa, Kenya.

29) Rietema, K., On the efficiency in separating mixtures of two components, Chemical Engineering Science, 7, 89, 1957.

30) Rood, M. J., Technological and economic evaluation of municipal solid waste incineration, OTT-2, Sept. 1988, University of Illinois Center for Solid Waste Management and Research office of Technology Transfer, Chicago, IL, 1988.

31) Salvato, J. A., Environmental Engineering and Sanitation, A Wiley-Interscience Publication, New York, 1982.

32) Schubeler, P., Wehrle, K. and Christen, J., Conceptual framework for municipal solid waste management in low-income countries, Urban Management and Infrastructure, UNDP, UNCHS (Habitat), World Bank, SDC Collaborative Program on Municipal Solid Waste Management in Low-Income Countries, August 1996, Working Paper No. 9,

SKAT (Swiss Centre for Development Cooperation in Technology and Management), Gallen, Switzerland, 1996.

33) Senate, E., Galtier, L., Bekaert, C., Lambolez-Michel, L. and Budka, A., Odor management at MSW landfill sites: odor sources, odorous compounds and control measures, Proceedings Sardinia, Ninth International Waste Management and Landfill Symposium, S. Marghorita di Pula, Cagliari, Italy, 6 – 10 October 2003.

34) Shah, K. L., Basics of solid and hazardous waste management technology, Prentice Hall; 1999.

35) Sonnemann, G., Castells, F., Schuhmacher, M., Integrated life-cycle and risk assessment for industrial processes, Lewis Publishers CRC Press Co., Boca Raton, 2004.

36) Stokoe J. and Teague, E., Integrated solid waste, USDA Rural Utilities Service, Washington, D.C., 1995.

37) Suess, M. J. ed., Solid waste management: Selected topics, WHO, Regional Office for Europe, Copenhagen, 1985.

38) Tchobanoglous, G., Theisen, H., and Eliassen, R., Solid waste engineering principles and management issues, McGraw-Hill Kogakusha, Ltd, Tokyo, 1977.

39) Tchobanoglous, G., Theisen H. and Vigil S., Integrated solid waste management: Engineering principles and management issues, Mc-Graw-Hill International Editions, New York, 1993

40) Tchobanoglous, G. and Kreith, F., Handbook of solid waste management, McGraw-Hill Professional; 2002.

41) Tedder, D. W., Pohland, F. G., Emerging technologies in hazardous waste management 8, Kluwer Academic Publishers, New York, 2002.

42) Twardowska, E., Edi., Solid waste: assessment, monitoring and remediation, Waste Management Series, Volume 4, ELSEVIER B.V., Amsterdam, 2004.

43) Vesilind, P. A., Worrell, W. A. and Reinhart, D. R., Solid waste engineering, Brooks/Cole, Thomson Learning, Bill Stenquist Pub., Pacific Grove, CA, 2002.

44) Van de Klundert, A. and Anschutz, J., Integrated sustainable waste management – The concept, WASTE, Gouda, The Netherlands, 2001.

45) Walsh, P. and O'Leary, P., Implementing municipal solid waste to energy systems, University of Wisconsin – Extension for Great Lakes Regional Biogas Energy Program, 1986.

46) Worrell, W. A. and Vesilind, P. A., Solid waste engineering, CL-Engineering Pub., 2012.

Appendixes

Appendix (1): Typical specific weight and moisture content data for residential, commercial, industrial, and agricultural wastes [2,4,6,8,10,23-25,46].

Type of waste	Specific weight, lb/yd³		Moisture content, % by weight	
	Range	Typical	Range	Typical
Residential (Compacted)				
Food wastes (mixed)	220 - 810	490	50 - 80	70
Paper	70 - 220	150	4 - 10	6
Cardboard	70 - 135	85	4 - 8	5
Plastics	70 - 220	110	1 - 4	2
Textiles	70 - 170	110	6 - 15	10
Rubber	170 - 340	220	1 - 4	2
Leather	170 - 440	270	8 - 12	10
Yard wastes	100 - 380	170	30 - 80	60
Wood	220 - 540	400	15 - 40	20
Glass	270 - 810	330	1 - 4	2
Tin cans	85 - 270	150	2 - 4	3
Aluminum	110 - 405	270	2 - 4	2
Other metal	220 - 1940	540	2 - 4	3
Dirt, ash, etc.	540 - 1685	810	6 - 12	8
Ashes	1095 - 1400	1255	6 - 12	6
Rubbish	150 - 305	220	5 - 20	15
Residential yard wastes				
Leaves (loose and dry)	50 - 250	100	20 - 40	30
Green grass (loose and moist)	350 - 500	400	40 - 80	60
Green grass (wet and compacted)	1000 - 1400	1000	50 - 90	80
Yard waste (shredded)	450 - 600	500	20 - 70	50
Yard waste (composted)	450 - 650	550	40 - 60	50
Municipal				
In compactor truck	300 - 760	500	15 - 40	20
In landfill				
Normally compacted	610 - 840	760	15 - 40	25
Well compacted	995 - 1250	1010	15 - 40	25
Commercial				
Food wastes (wet)	800 - 1600	910	50 - 80	70
Appliances	250 - 340	305	0 - 2	1
Wooden crates	185 - 270	185	10 - 30	20
Tree trimmings	170 - 305	250	20 - 80	5
Rubbish (combustible)	85 - 305	200	10 - 30	15
Rubbish (noncombustible)	305 - 610	505	5 - 15	10
Rubbish (mixed)	235-305	270	10-25	15
Construction and demolition				
Mixed demolition (noncombustible)	1685 -2695	2395	2-10	4
Mixed demolition (combustible)	505 -675	605	4-15	8
Mixed construction (combustible)	305 -605	440	4-15	8
Broken concrete	2020 -3035	2595	0- 5	-
Industrial				
Chemical sludges (wet)	1350 - 1855	1685	75 - 99	80
Fly ash	1180 - 1515	1350	2 - 10	4
Leather scraps	170 - 420	270	6 - 15	10
Metal scrap (heavy)	2530 - 3370	3000	0 - 5	-
Metal scrap (light)	840 - 1515	1245	0 - 5	-

Metal scrap (mixed)	1180 - 2530	1515	0 - 5	-
Oils, tars, asphalts	1350 - 1685	1600	0 - 5	2
Sawdust	170 - 590	490	10 - 40	20
Textile wastes	170 - 370	305	6 - 15	10
Wood (mixed)	675 - 1140	840	30 - 60	25
Agricultural				
Agricultural (mixed)	675 - 1265	945	40 - 80	50
Dead animals	340 - 840	605	-	-
Fruit wastes (mixed)	420 - 1265	605	60 - 90	75
Manure (wet)	1515 - 1770	1685	75 - 96	94
Vegetable wastes (mixed)	340 - 1180	605	60 - 90	75

174

IA																VIIA	VIIIA
1 H 1.00794	IIA											IIIA	IVA	VA	VIA	1 H 1.00794	2 He 4.002602
3 Li 6.941	4 Be 9.012182											5 B 10.811	6 C 12.0107	7 N 14.00674	8 O 15.9994	9 F 18.9984032	10 Ne 20.1797
11 Na 22.989770	12 Mg 24.3050	IIIB	IVB	VB	VIB	VIIB		VIIIB			IB	IIB	13 Al 26.981538	14 Si 28.0855	15 P 30.973761	16 S 32.066	17 Cl 35.4527	18 Ar 39.948
19 K 39.0983	20 Ca 40.078	21 Sc 44.955910	22 Ti 47.867	23 V 50.9415	24 Cr 51.9961	25 Mn 54.938049	26 Fe 55.845	27 Co 58.933200	28 Ni 58.6934	29 Cu 63.546	30 Zn 65.39	31 Ga 69.723	32 Ge 72.61	33 As 74.92160	34 Se 78.96	35 Br 79.904	36 Kr 83.80
37 Rb 85.4678	38 Sr 87.62	39 Y 88.90585	40 Zr 91.224	41 Nb 92.90638	42 Mo 95.94	43 Tc (98)	44 Ru 101.07	45 Rh 102.90550	46 Pd 106.42	47 Ag 107.8682	48 Cd 112.411	49 In 114.818	50 Sn 118.710	51 Sb 121.760	52 Te 127.60	53 I 126.90447	54 Xe 131.29
55 Cs 132.90545	56 Ba 137.327	57 La* 138.9055	72 Hf 178.49	73 Ta 180.9479	74 W 183.84	75 Re 186.207	76 Os 190.23	77 Ir 192.217	78 Pt 195.078	79 Au 196.96655	80 Hg 200.59	81 Tl 204.3833	82 Pb 207.2	83 Bi 208.98038	84 Po (209)	85 At (210)	86 Rn (222)
87 Fr (223)	88 Ra (226)	89 Ac** (227)	104 Rf (261)	105 Db (262)	106 Sg (263)	107 Bh (262)	108 Hs (265)	109 Mt (266)	110 Ds (269)	111 Uuu (272)	112 Uub (277)		114 Uuq (289) (287)		116 Uuh (289)		118 Uuo (293)

* Lanthanide series	58 Ce 140.116	59 Pr 140.90765	60 Nd 144.24	61 Pm (145)	62 Sm 150.36	63 Eu 151.964	64 Gd 157.25	65 Tb 158.92534	66 Dy 162.50	67 Ho 164.93032	68 Er 167.26	69 Tm 168.93421	70 Yb 173.04	71 Lu 174.967
** Actinide series	90 Th 232.0381	91 Pa 231.03588	92 U 238.0289	93 Np (237)	94 Pu (244)	95 Am (243)	96 Cm (247)	97 Bk (247)	98 Cf (251)	99 Es (252)	100 Fm (257)	101 Md (258)	102 No (259)	103 Lr (262)

Appendix (2): Periodic Table of Elements.

Appendix 3
Screenshots from the listed computer programs

Example 1.1 – Form1.vb (Design):

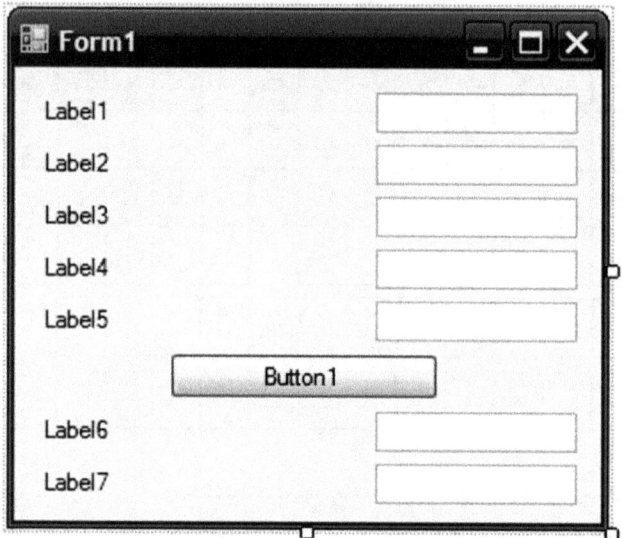

Example 1.1 – Running window:

Mixed house waste	
Recyclables	
Commercial waste	
Construction and demolition debris	
Leaves and miscellaneous	

Calculate

| Diversion (% of MSW) | |
| Diversion (% of refuse) | |

Example 2.1 – Form1.vb (Design):

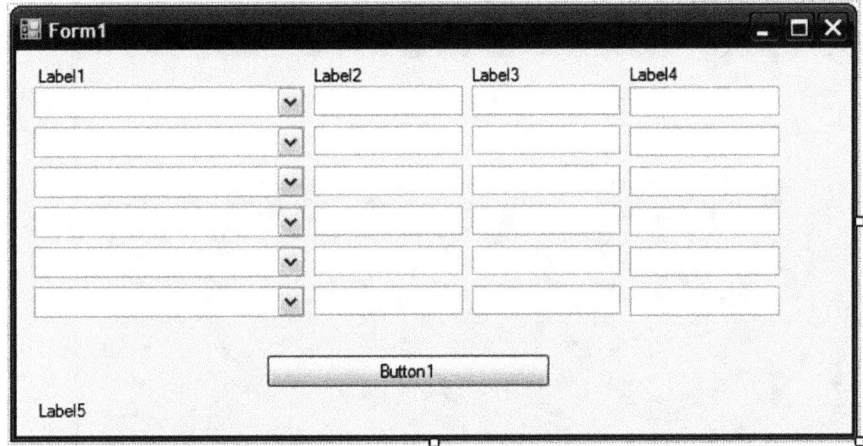

Example 2.1 – Running window:

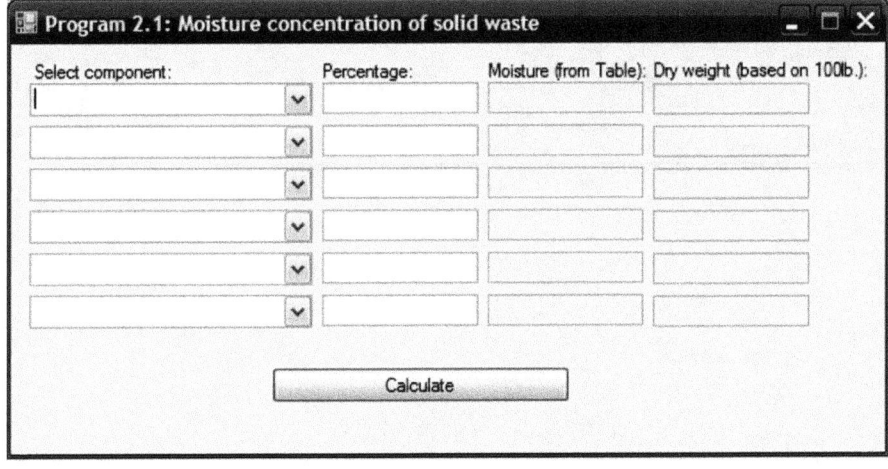

Example 2.2 – Form1.vb (Design):

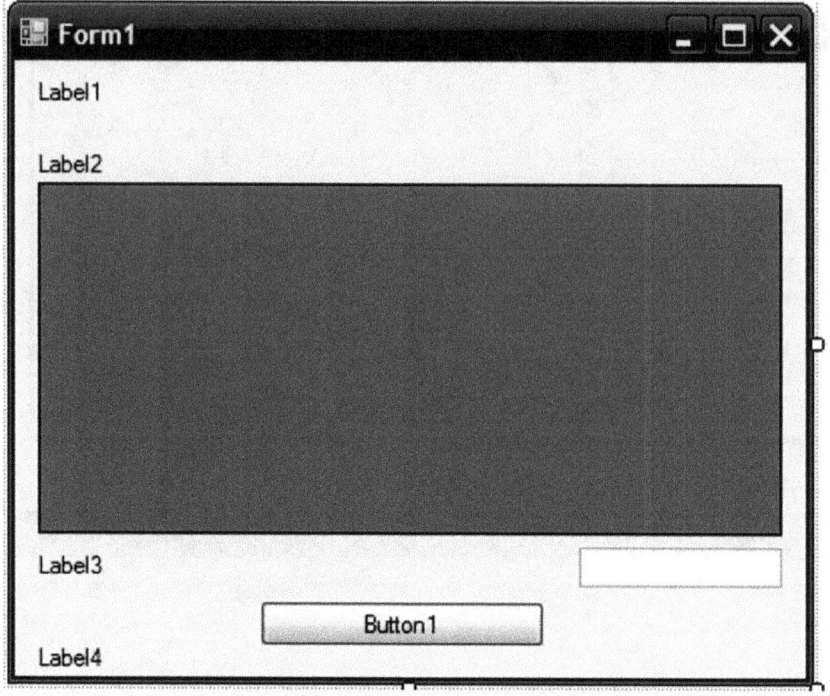

Example 2.2 – Running window:

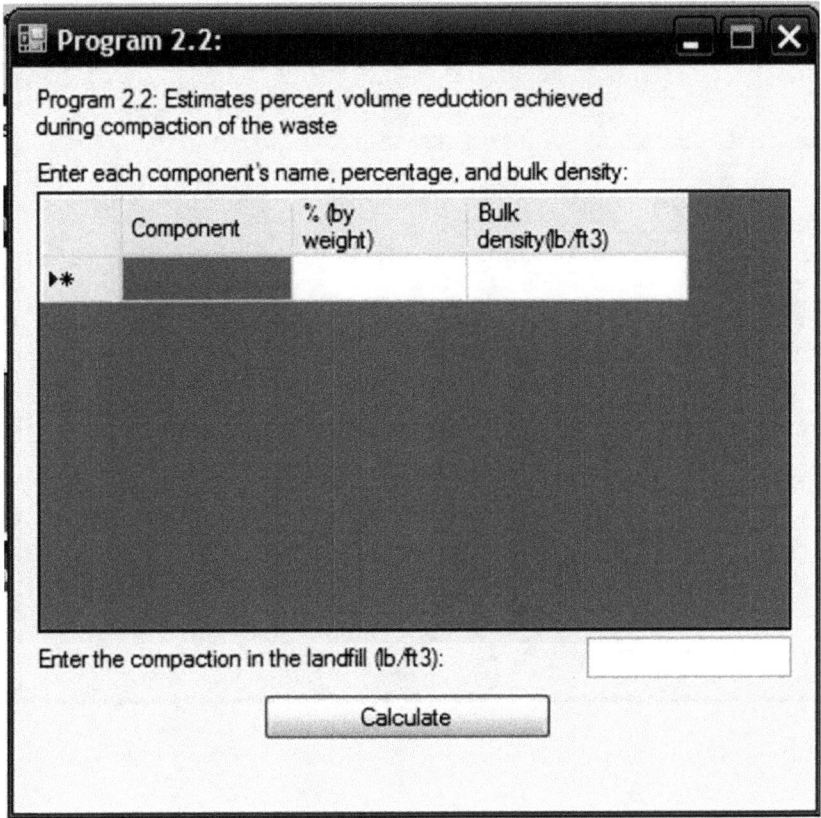

Example 2.3 – Form1.vb (Design):

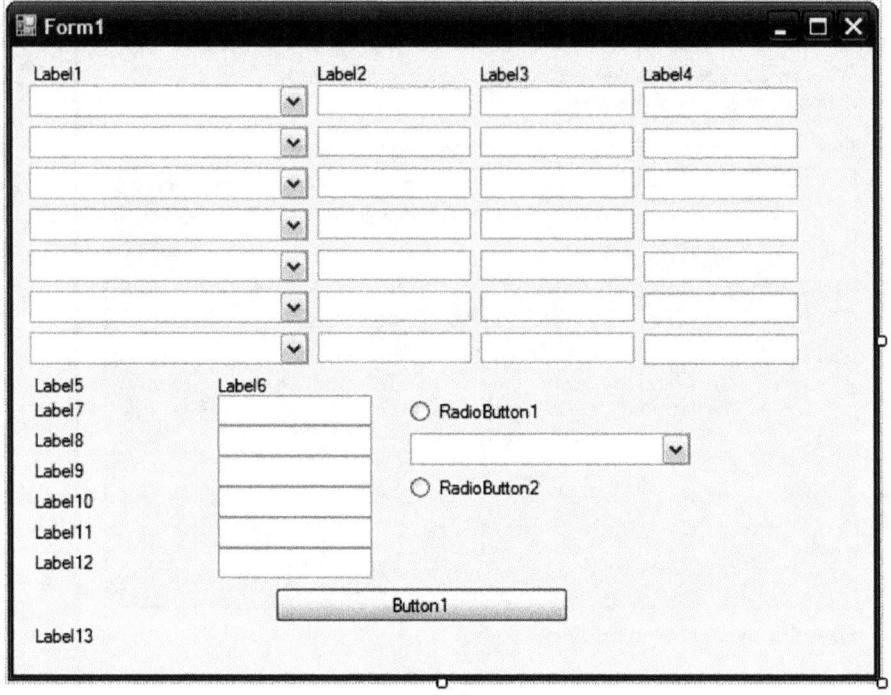

Example 2.3 – Running window:

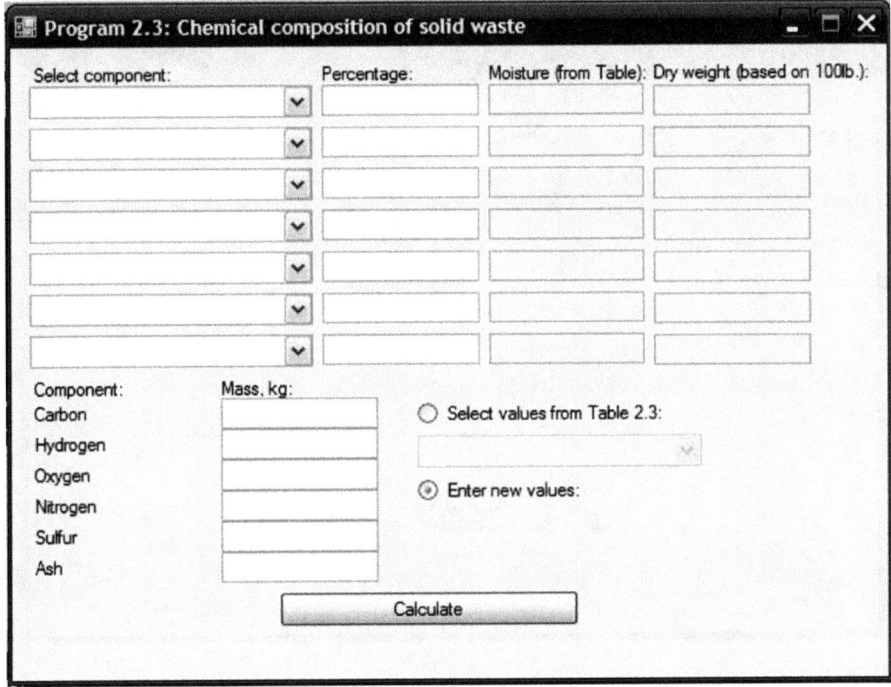

Example 2.4 – Form1.vb (Design):

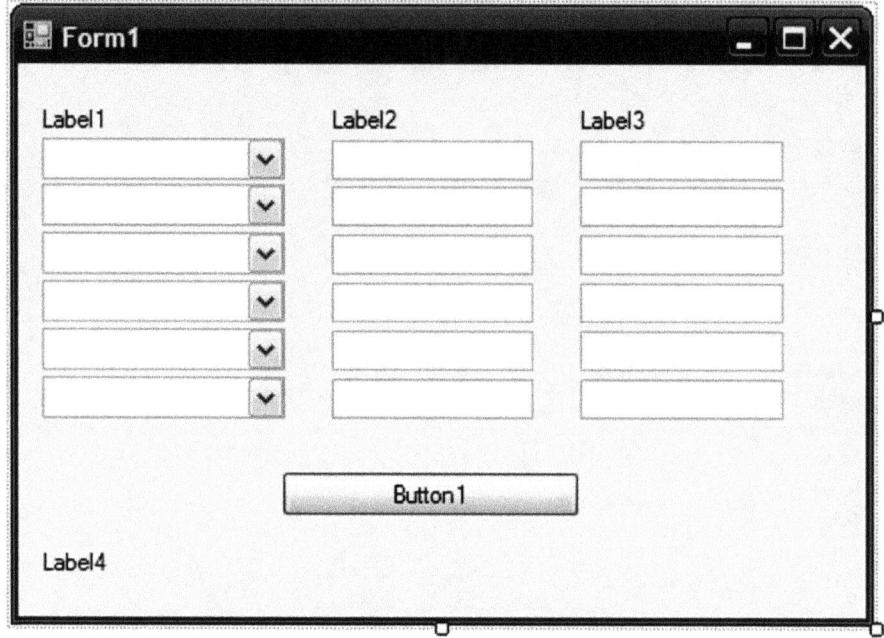

Example 2.4 – Running window:

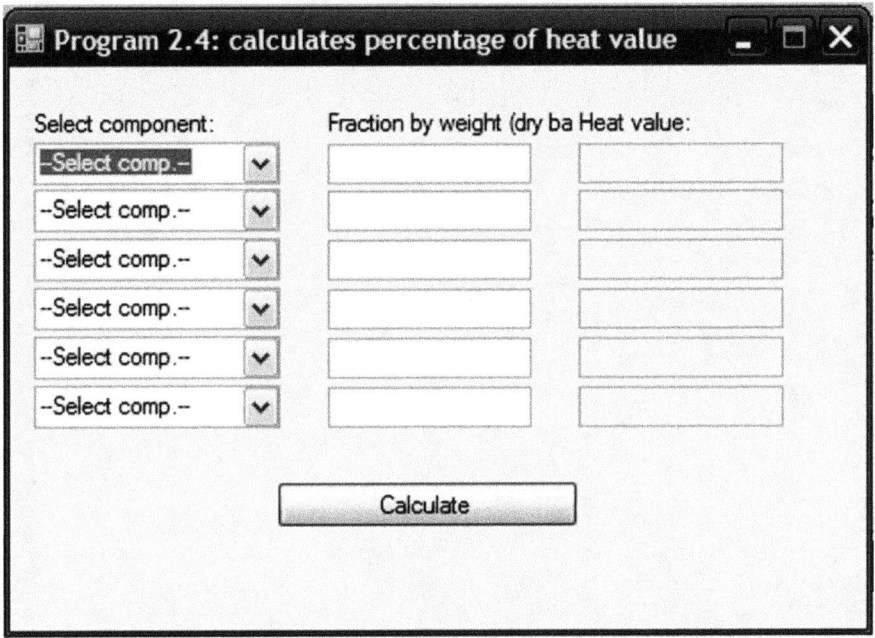

Example 3.1 – Form1.vb (Design):

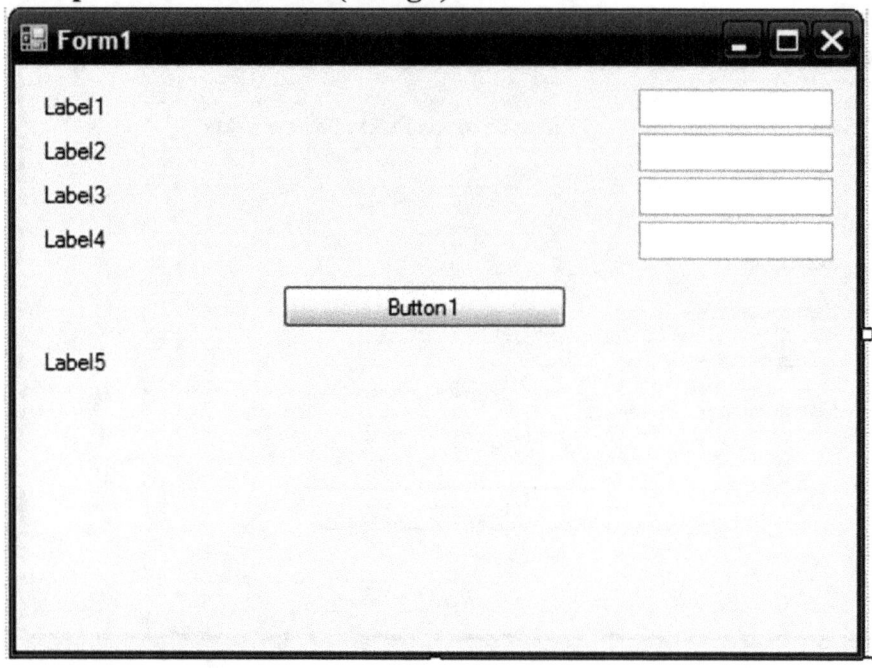

Example 3.1 – Running window:

Program 3.1:Calculates number of cans and compac...

Number of people in the family:

Rate of daily generation of solid waste (lb./capita/day):

Bulk density of refuse in a typical garbage can (lb/yd3):

Collection frequency (per week):

Calculate

Example 3.2 – Form1.vb (Design):

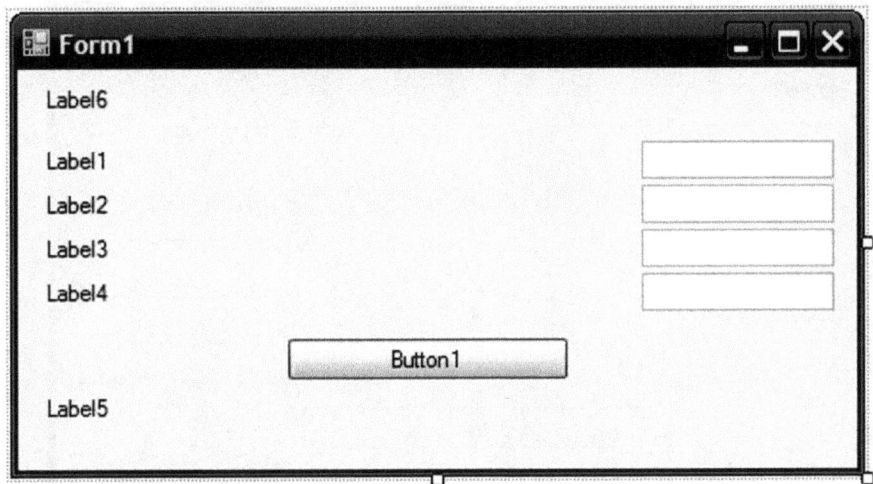

Example 3.2 – Running window:

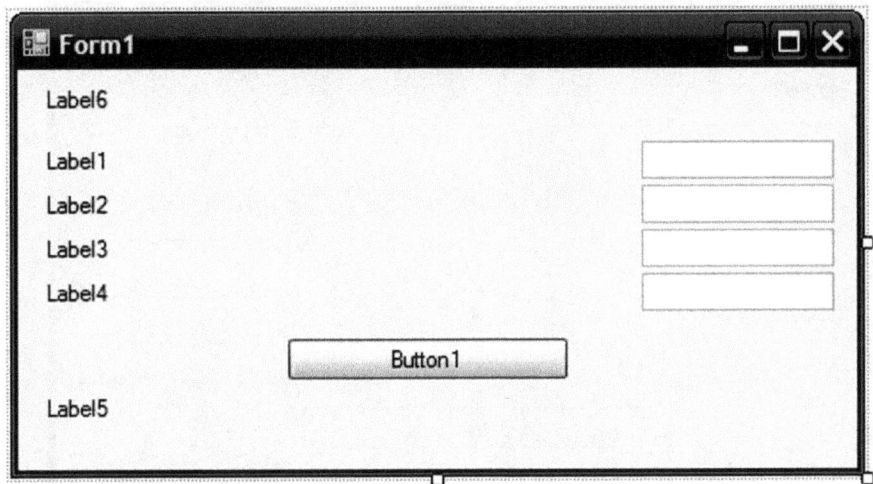

Example 4.1 – Form1.vb (Design):

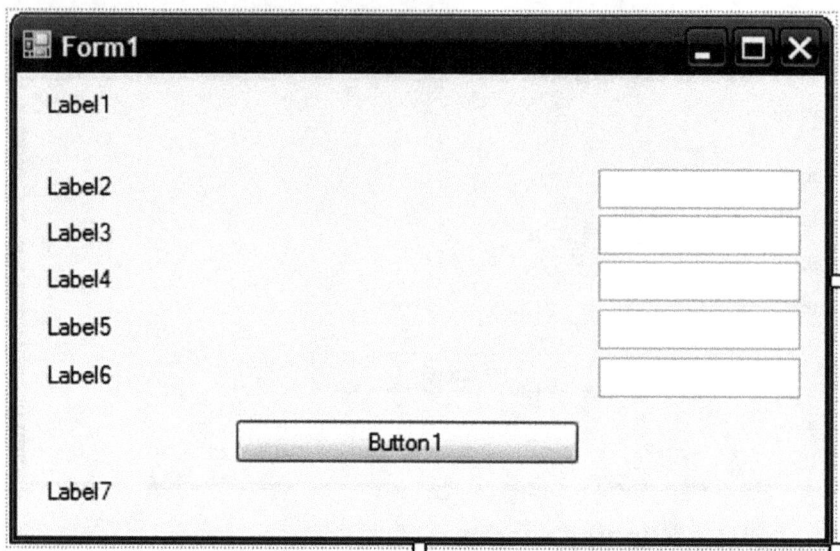

Example 4.1 – Running window:

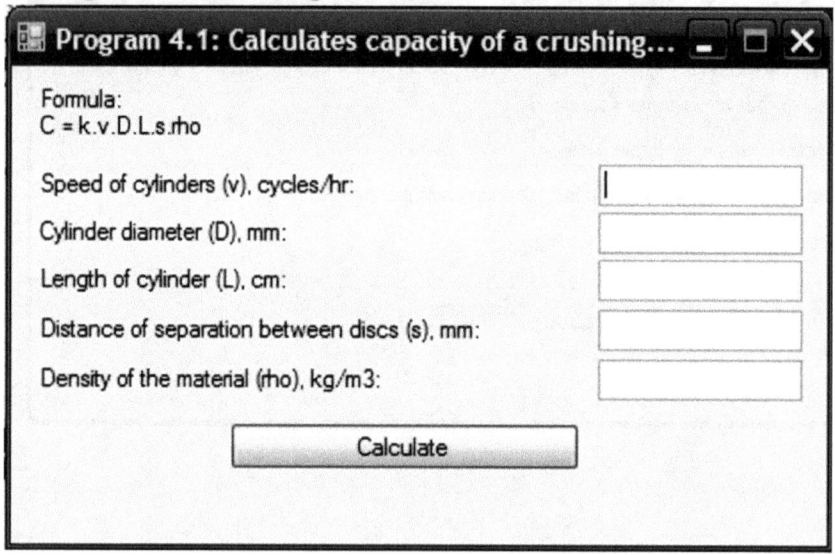

Example 4.2 – Form1.vb (Design):

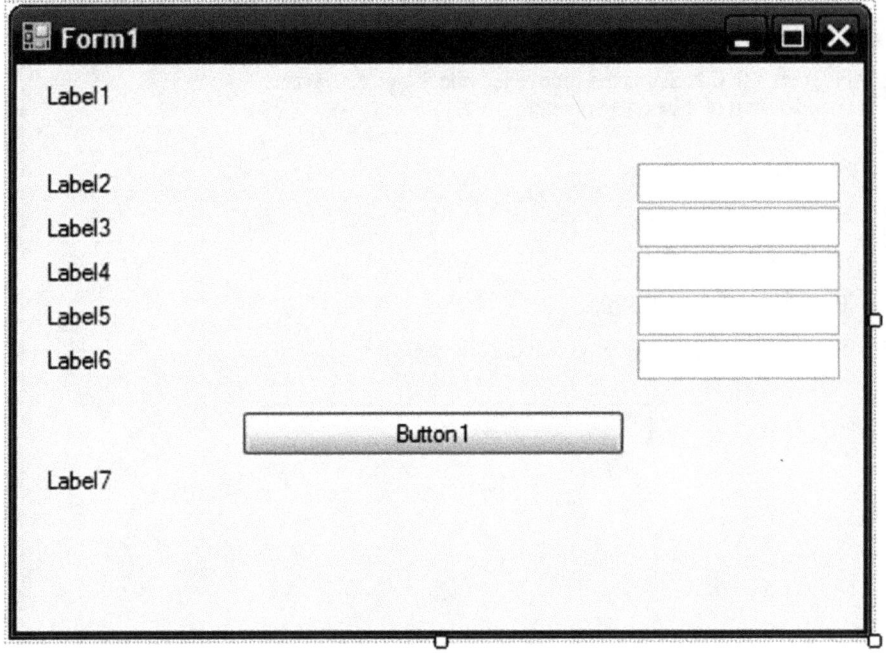

Example 4.2 – Running window:

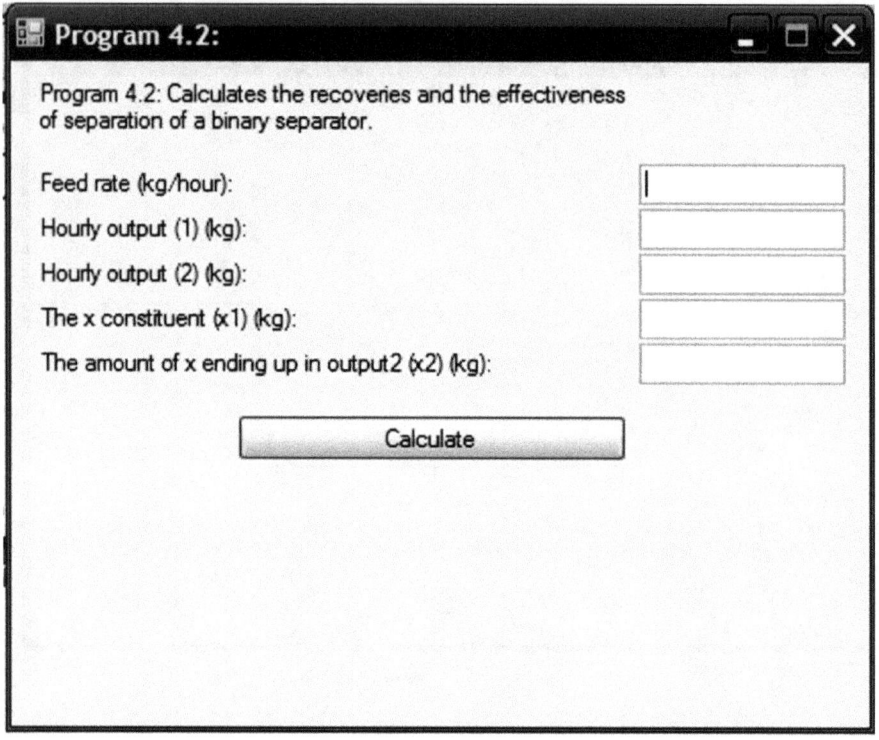

Example 5.1 – Form1.vb (Design):

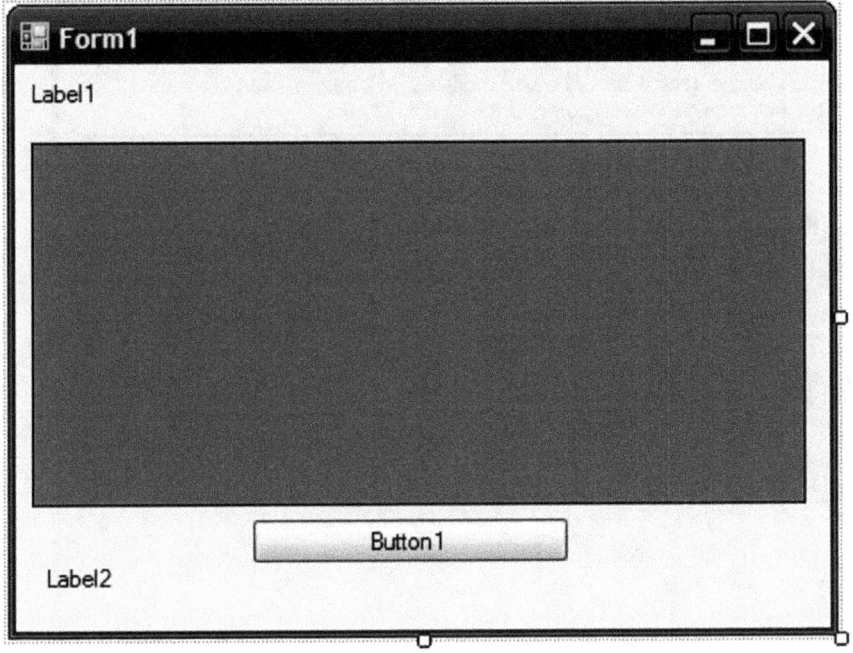

Example 5.1 – Running window:

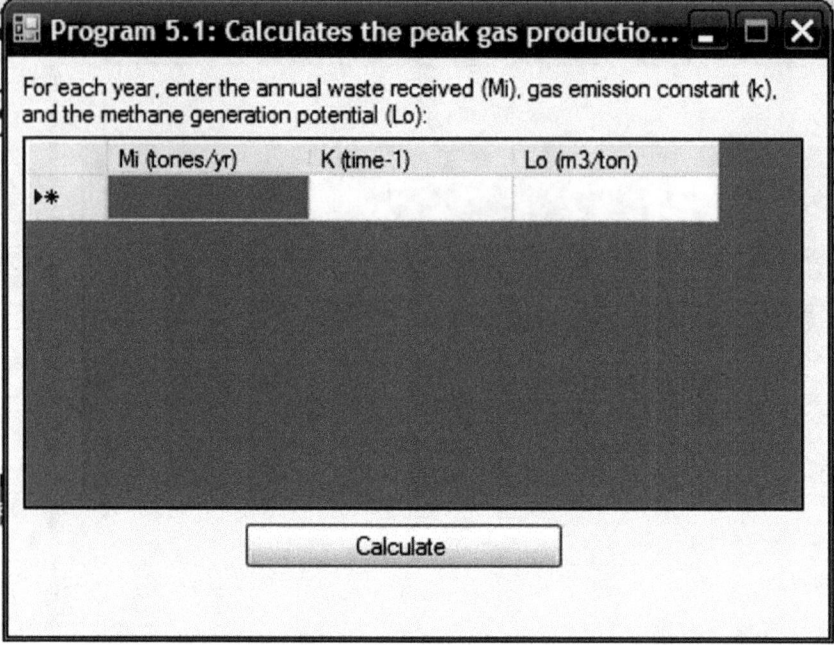

Example 6.1 – Form1.vb (Design):

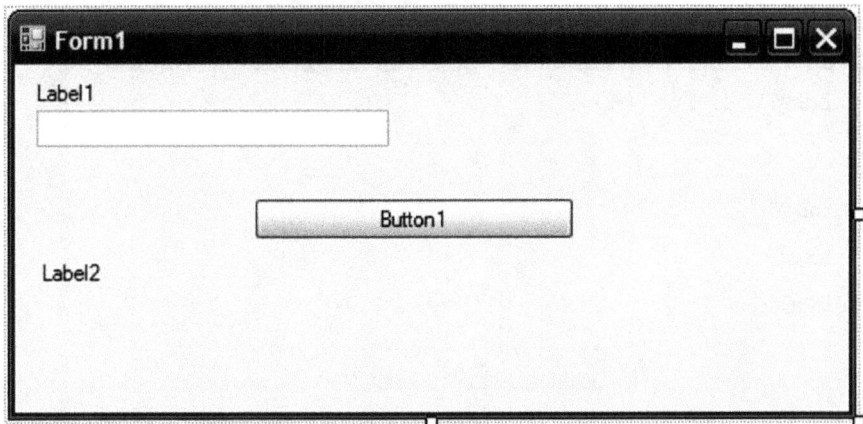

Example 6.1 – Running window:

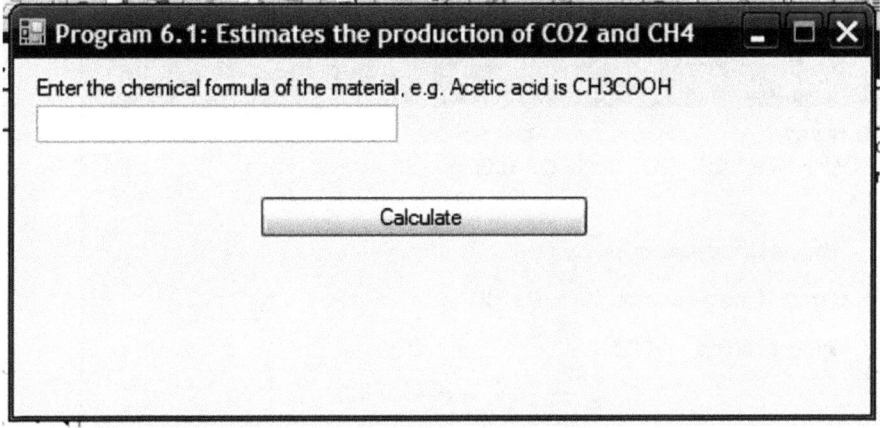

Example 6.2 – Form1.vb (Design):

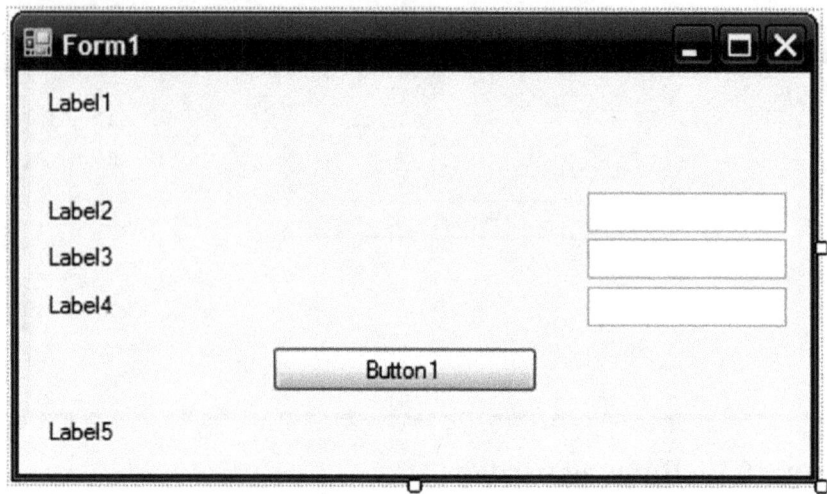

Example 6.2 – Running window:

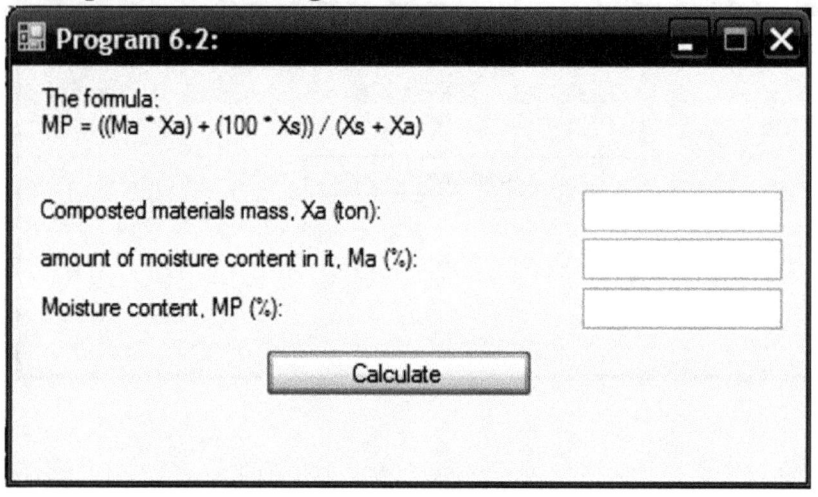

Example 6.3 – Form1.vb (Design):

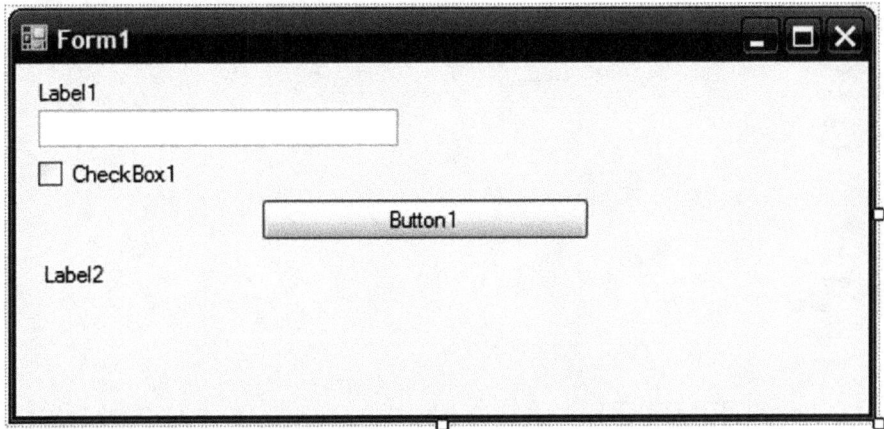

Example 6.3 – Running window:

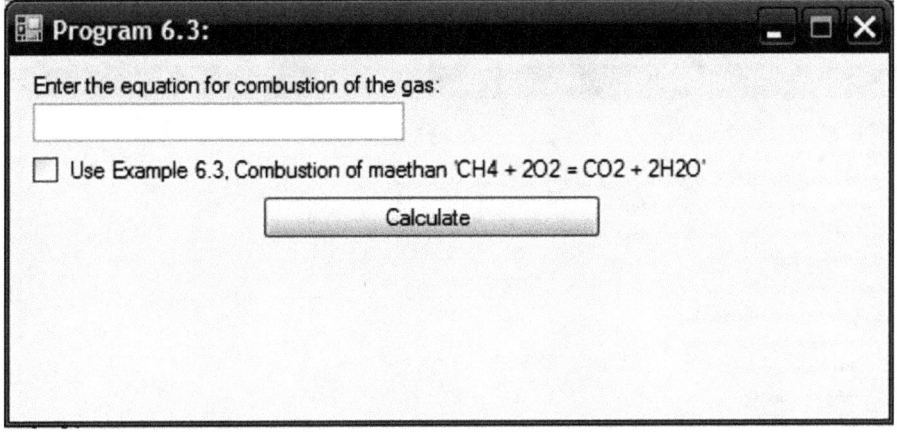

Example 6.4 – Form1.vb (Design):

Example 6.4 – Running window:

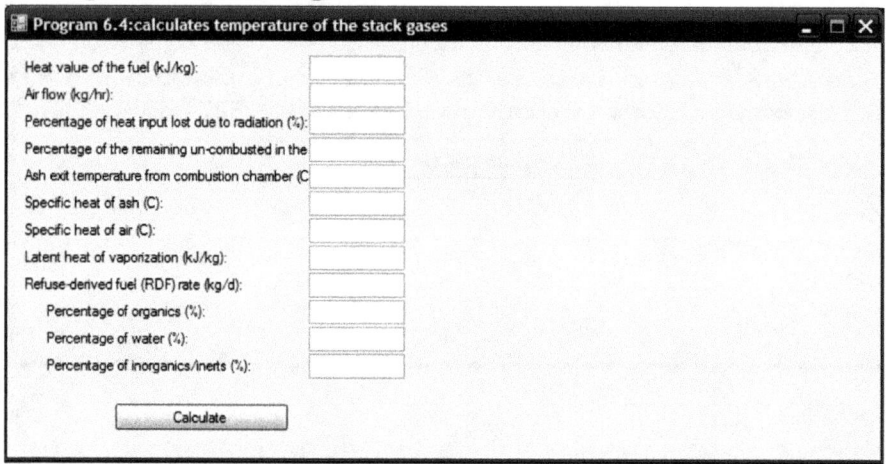

Example 7.1 – Form1.vb (Design):

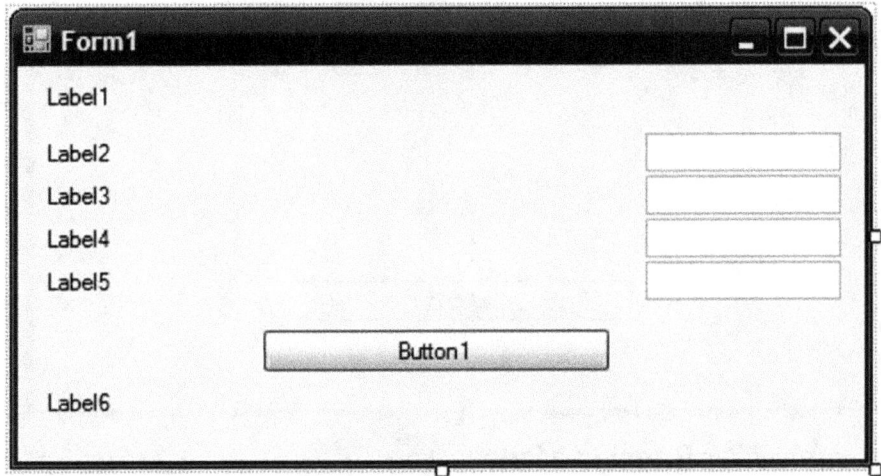

Example 7.1 – Running window:

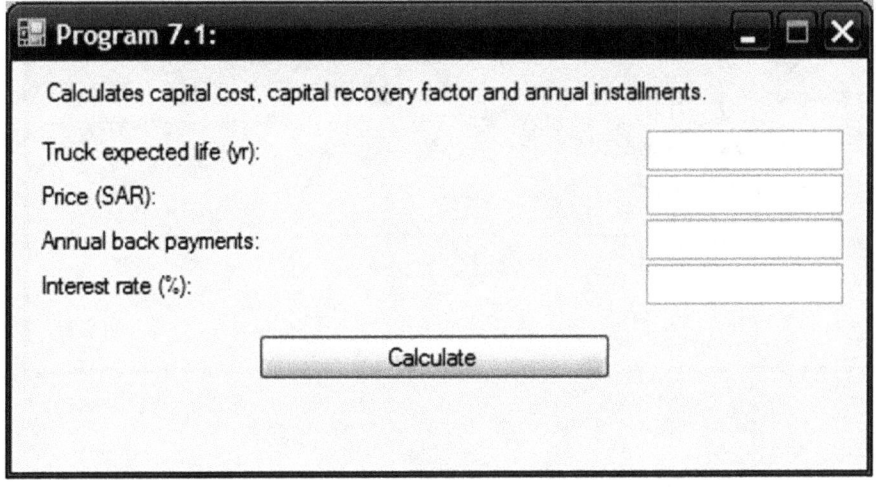

Example 7.2 – Form1.vb (Design):

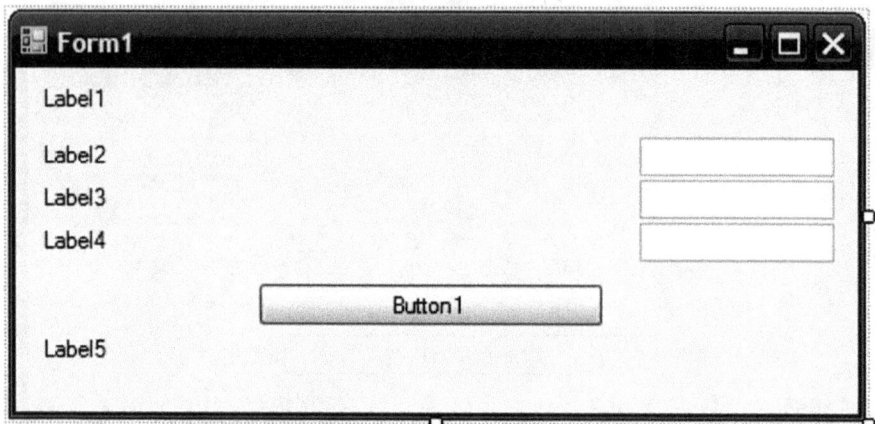

Example 7.2 – Running window:

Example 7.3 – Form1.vb (Design):

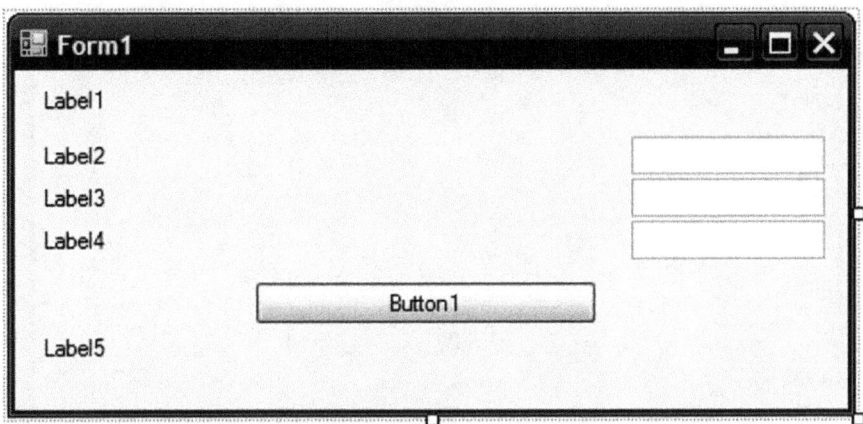

Example 7.3 – Running window:

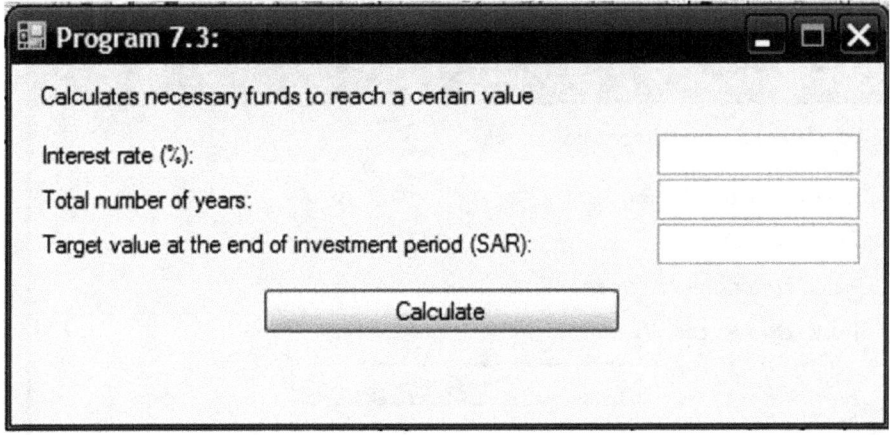

Example 7.4 – Form1.vb (Design):

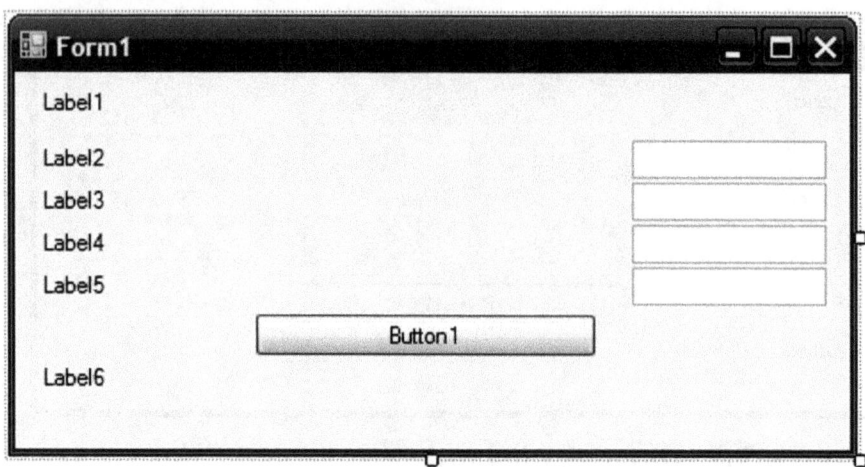

Example 7.4 – Running window:

Program 7.4:

Calculates refuse collection truck's costs

Expected life of collection truck:

Truck cost (SAR):

Annual installments interest rate (%):

Truck operating cost:

Calculate

Answers to practical exercises

Chapter one: Exercise (1.2)
1) 32%, 9%

Chapter two: Exercise (2.2)
1) 69.5%
2) 25.1 %
3) 30.8 %
4) 143 lb/yd^3, 0.11; 232 lb/yd^3, 0.18
5) 86 kg/m^3, 0.12; 11 kg/m^3, 0.15
6) 21 %
7) 95 kg/m^3, 0.14; 250 kg/m^3, 0.36
8) 24 kg/m^3, 0.03; 18 kg/m^3, 0.03
9) Chemical formula with sulfur $C_{934.4}H_{2489.6}O_{1004.8}N_{11.5}S$; Chemical formula without sulfur: $C_{81.1}H_{216.1}O_{87.2}N$; 14125 Btu/lb
10) Chemical formula with sulfur is: $C_{860}H_{2513.3}O_{1050}N_{14.3}S$; Chemical formula without sulfur is: $C_{60}H_{175.3}O_{73.86}N$; 1162 Btu/lb
11) 5065 Btu/lb = 11782 kJ/kg
12) 5772 Btu/lb
13) Chemical formulas without sulfur:
    1. without water $C_{60}H_{94.3}O_{37.8}N$
    2. with water $C_{60}H_{156.3}O_{69.1}N$
    Chemical formulas with sulfur:
    1. without water $C_{760}H_{1194.7}O_{478.7}N_{12.7}S$
    2. with water $C_{760}H_{1980}O_{874.7}N_{12.7}S$
14) 30%, Chemical formula with sulfur is: $C_{548.8}H_{1612.8}O_{637.9}N_{8.83}S$, Chemical formula without sulfur is: $C_{62.1}H_{182.5}O_{72.2}N$, 1749 Btu/lb
15) 25.5%; chemical formula with sulfur is: $C_{551.1}H_{1521.1}O_{588.4}N_{7.6}S$; Chemical formula without sulfur is: $C_{72.3}H_{199.6}O_{77.2}N$; 14487 kJ/kg
16) 5172 Btu/lb

17) 9091 Btu/lb

18) 171 cal/°C

19) 10471 Btu/lb

Chapter three: Exercise (3.2)

1) 2, one 30-gal can
2) 143 customer
3) three 30-gallon cans, four 20-lb blocks, one 30-gal can
4) 240 customer
5) 184 customer
6) 720 customer
7) 480 customer
8) 900 customer
9) 27 cars/hr
10) 600 customers/d
11) 450 customers/d
12) 4 vehicles
13) 5 vehicles
14) 3.3 tones/hr

Chapter four: Exercise (4.2)

1) 4.14 tones/hr
2) 6.75 tones/hr
3) 89%, 94%, 58% 78%

Chapter five: Exercise (5)

1) 552 mm/yr
2) 280 mm/y
3) 705 mm/yr
4) 1572122 m³
5) 11 cm
6) 15 m

Chapter six: Exercise (6.2)

1) $C_6 H_{12} O_6$ = $3CH_4 + 3CO_2$; 0.27 kg CH_4 and 0.73 kg CO_2, 67.2 liters
2) 0.37 kg CH_4; 0.88 kg CO_2
3) 16.5 tonnes of water or sludge to be added
4) $C_2H_5OH = 1.5CH_4 + 0.5CO_2$ or $2C_2H_5OH = 3CH_4 + CO_2$
   $CH_3CH_2COOH + 0.5H_2O = 1.75CH_4 + 1.25CO_2$ or
   $4CH_3CH_2COOH + 2H_2O = 7CH_4 + 5CO_2$
   $C_4H_8O_2 + H_2O = 2.5CH_4 + 1.5CO_2$ or $2C_4H_8O_2 + 2H_2O = 5CH_4 + 3CO_2$; 1.2, 1.8 kg; 201, 336 litres
5) 41%
6) 1238 °C, 2260 °F
7) 572 °C, 1062 °F

Chapter seven: Exercise (7)

1) SAR 72,664
2) SAR 86,310
3) 6.131%
4) SAR 662,098
5) SAR 134,136
6) SAR 72,664

The authors at a glance

Prof. Dr. Eng. **Isam Mohammed Abdel-Magid Ahmed**, B.Sc., PDH, DSE, Ph.D., FSES, CSEC, MSECS. Professor of Water Resources and Environmental Engineering.
Place & date of birth: Rufa'a, 1952

Education & experience: B.Sc., Honors (first class) U. Khartoum 1977, Diploma Hydrology, U. Padova (Italy) 1978, M.Sc. Delft U. Tech. (The Netherlands) 1979, Ph.D., U. Strathclyde (GB) 1982.

Professional Experience: Civil engineer General Corporation for Irrigation and Drainage-Sudan, U. Khartoum, U. United Arab Emirates, U. Sultan Qaboos, Omdurman Islamic U., Sudan U. Sci. & Tech., Juba U., Industrial Research & Consultancy Center, Sudan Academy for Sciences, King Faisal U., U. Dammam. Supervised many Diploma, B.Sc., M.Sc., & Ph.D. work. External examiner to different institutions. Founder of some institutions, ccenters & refereed journals.

Awards: Sudan Engineering Society Prize for the Best Project in Civil Eng., MoI Prize for Second Best Performance in 4th year Civil Engineering, Honourly Scarf for Enrichment of Knowledge, UoK Press. Best book of the year - MoIC Sudanese Press Council, ALECSO prize for a book in engineering.

Publications: Authored or co-authored over many papers, publications, text and reference books, technical reports, lecture notes in areas of water supply; wastewater disposal, reuse & reclamation; solid waste disposal; water resources.

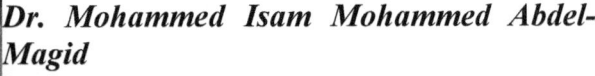

Dr. Mohammed Isam Mohammed Abdel-Magid

Dr. Mohammed Isam Mohammed Abdel-Magid (MBBS, BLS, ALS, MRCP-UK Part I&II Written) is a graduate of the College of Medicine, University of Khartoum, Sudan, 2008. He completed basic training with the Ministry of Health, Sudan, then worked as a physician in the department of Internal Medicine, Ribat University hospital, Sudan, and the Ministry of Health, Kingdom of Saudi Arabia.

He is completing his higher training with the membership of the Royal Colleges of Physicians of the United Kingdom (MRCP-UK) of which he completed its three parts.

He tutored in problem-based learning teaching sessions in the department of Internal Medicine, Sudan International University, Sudan.

He is a registered practicing physician with the Sudan Medical Council, the Health Authority of Abu-Dhabi (HAAD), and the Saudi Commission of Health Specialties (SCHS). He is a full member of the Society of Acute Medicine of UK (SAM), the European Society for Emergency Medicine (EuSEM), and the European Respiratory Society (ERS).

He is a peer reviewer with the Science Journal of Medicine & Clinical Trial and the Pan-African Journal of Medical Sciences.

www.ingramcontent.com/pod-product-compliance
Lightning Source LLC
Chambersburg PA
CBHW051906170526
45168CB00001B/267